WHAT ARE YOU OPTIMISTIC ABOUT?

BOOKS BY JOHN BROCKMAN

As author:
By the Late John Brockman
37
Afterwords
The Third Culture: Beyond the Scientific Revolution
Digerati

As editor:
About Bateson
Speculations
Doing Science
Ways of Knowing
Creativity
The Greatest Inventions of the Past 2,000 Years
The Next Fifty Years
The New Humanists
Curious Minds
What We Believe But Cannot Prove
My Einstein
Intelligent Thought
What Is Your Dangerous Idea?

As coeditor:
How Things Are

WHAT ARE YOU OPTIMISTIC ABOUT?

Today's Leading Thinkers
on Why Things Are Good and Getting Better

edited by JOHN BROCKMAN

HARPER PERENNIAL

NEW YORK • LONDON • TORONTO • SYDNEY

HARPER ● PERENNIAL

FIRST U.S. EDITION

Library of Congress Cataloging-in-Publication Data has been applied for.

ISBN: 978-0-06-143693-2
ISBN-10: 0-06-143693-3

07 08 09 10 11 RRD 10 9 8 7 6 5 4 3 2 1

WHAT ARE YOU OPTIMISTIC ABOUT?

CONTENTS

PREFACE : THE ANNUAL *EDGE* QUESTION

In 1991, I suggested the idea of a third culture, which 'consists of those scientists and other thinkers in the empirical world who, through their work and expository writing, are taking the place of the traditional intellectual in rendering visible the deeper meanings of our lives, redefining who and what we are.' By 1997, the growth of the Internet had allowed implementation of a home for the third culture on the Web, on a site named *Edge* (www.edge.org).

Edge is a celebration of the ideas of the third culture, an exhibition of this new community of intellectuals in action. They present their work, their ideas, and comment about the work and ideas of third-culture thinkers. They do so with the understanding that they are to be challenged. What emerges is rigorous discussion concerning crucial issues of the digital age in a highly charged atmosphere where 'thinking smart' prevails over the anesthesiology of wisdom.

The ideas presented on *Edge* are speculative; they represent the frontiers in such areas as evolutionary biology, genetics, computer science, neurophysiology, psychology, and physics. Some of the fundamental questions posed are: Where did the universe come from? Where did life come from? Where did the mind come from? Emerging out of the third culture is a new natural philosophy, new ways of understanding physical systems, new ways of thinking that call into question many of our basic assumptions of who we are, of what it means to be human.

An annual feature of *Edge* is The World Question Center, which was introduced in 1971 as a conceptual art project by my friend and collaborator the late artist James Lee Byars. His plan was to gather the hundred most brilliant minds in the world together in a room, lock them in, and 'have them ask each other the questions they were asking themselves.' The result was to be a synthesis of all thought. Between idea and execution, however, are many pitfalls. Byars identified his hundred most brilliant minds, called each of them, and asked them what questions they were asking themselves. The result: seventy people hung up on him.

But by 1997, the Internet and e-mail had allowed for a serious implementation of Byars' grand design, and this resulted in launching *Edge*. For each of the anniversary editions of *Edge*, I have used the interrogative myself and asked contributors for their responses to a question that comes to me or one of my correspondents in the middle of the night. Here is the 2007 *Edge* Question:

As an activity, as a state of mind, science is fundamentally optimistic. Science figures out how things work and thus can make them work better. Much of the news is either good news or news that can be made good, thanks to ever deepening knowledge and ever more efficient and powerful tools and techniques. Science, on its frontiers, poses more and ever better questions, ever better put.

What are you optimistic about? Why? Surprise us!

John Brockman
Publisher & Editor, *Edge*

ACKNOWLEDGMENTS

I wish to thank John Williams of HarperCollins, U.S., and Andrew Gordon of The Free Press, U.K., for their encouragement.

I am also indebted to my agent, Max Brockman, who recognized the potential for this book, and to Sara Lippincott for her thoughtful and meticulous editing.

INTRODUCTION BY DANIEL C. DENNETT

I hear the human race
Is fallin' on its face,
And hasn't very far to go.

So sings naïve Nellie Forbush in *South Pacific*. This self-styled 'cock-eyed optimist' is quite aware that she's bucking the fashion, the centuries-long tradition of disdain for those whose outlook is too rosy. Along with Pollyanna and Dr. Pangloss, she symbolizes the top-down, fact-free ideology of progress: Things will get better; we can do it! We're living in the best of all possible worlds! Nellie knows that the word on the street is that this is folly, but she's 'stuck like a dope with a thing called hope.'

The contributors to this very cheering anthology are also full of hope, but theirs is a different brand of optimism, born of expertise and hard, imaginative thinking. And one of the most optimistic things about the collection is the breadth and variety of things the contributors are optimistic about. So many different ways we can make the world better! So many lights at the end of so many tunnels! Here we find schemes for cooling the Arctic ice cap, solving our energy problems, democratizing the global economy, improving transparency in government, muffling or dissolving religious discord, and even enlarging our personal intelligence and improving the phenomenon of friendship. We can come to understand ourselves and each other better, finally master math, and share our good fortune with larger segments of the world's population (which will soon stabilize).

It's all too good to be true, of course. That is, it can't *all* be sound prognostication. Some of the schemes will eventually prove to be cock-eyed, but we can't tell which ones until we try them and test them. This is part of the strength of the phenomenon: We have an open forum of candidates that can compete for credibility and feasibility, and the competition – if we manage it right – will be judged on excellence, not political support or authoritarian fiat. It's *not* who you know; it's *what* you know.

Knowledge is the thread that runs through all the entries. Not Knowledge of the (Divine, Mysterious) Truth, but good old knowledge of facts, (lower-case) truths dug up and confirmed by careful testing – the kind of knowledge that has been steadily accumulating in the human race for thousands of years and is now expanding explosively on almost all topics. With some few remarkable – and much analyzed – exceptions, once we human beings figure something out, it stays figured out. We can add it to our capacious treasurehouse of facts and put it to a hundred different uses, as opportunities arise. This store of shared and transmitted knowledge is surely what explains the enormous change that the human race has brought about on our planet in recent times. Ten thousand years ago, long after the taming of fire but not so long after the dawn of agriculture, our ancestors, together with all their livestock and pets, didn't make much of a dent – approximately 0.1 of 1 percent of the terrestrial biomass, according to calculations made by Paul MacCready, the green-minded chairman of Aerovironment, Inc. We were just another primate with some unique habits.

Today, we and our domesticated animals make up about 98 percent of the terrestrial biomass, and what little wildlife remains is heavily dependent on us for its survival. As MacCready put it in a recent Web essay ('An Ambivalent Luddite at a Techonological Feast'):

> Over billions of years, on a unique sphere, chance has painted a thin covering of life – complex, improbable, wonderful, and fragile. Suddenly we humans (a recently arrived species no longer subject to the checks and balances inherent in nature), have grown in population, technology, and intelligence to a position of terrible power: we now wield the paintbrush.

The explosive growth of power – of sheer *can do* – enjoyed by our human species means that many things are possible now that were never before possible, for good and for ill. Is there any reason to suppose that we will use this power mainly for good? In fact, there is – even in the face of all the 'man-made' calamities this power has enabled. This comes out vividly in many of the contributions. First, however myopic our initial forays into a project or campaign may be, there is a built-in tradition – across the spectrum of human activity – of reflection and criticism, a habit of reevaluation that is designed to look for flaws and problems, opportunities to improve, signs that a midcourse correction is called for. Science sets the gold standard for this, with its layers and layers of review, evaluation, testing, and policing. It is no coincidence that when sciencephobes want to brandish examples of scientific folly, prejudice, misrepresentation, and sheer dishonesty, they invariably help themselves to the results of investigations conducted by the sciences themselves.

Why? Because no other institution is better equipped for such intense self-policing, such relentless self-exposure. Indeed, when other institutions clean house – when the media set out to investigate their own biases and shortcomings, or when businesses or churches or government institutions engage in self-evaluation – they use techniques of unbiased data-gathering, statistics, experimentation, and argumentation that have been honed by the sciences over the centuries. And everywhere you look, you find people 'going meta-,' adding a recursive loop of inquiry on top of the inquiries they have made so far. For every analysis, there is meta-analysis. There are not just composers and musicians, there are music critics and prize committees – and pundits ready to issue judgments about the quality of the work of those critics and committees. There are people who earn a good living selling information about whom to ask about whom to ask about which stocks to purchase. Need to hire a new consultant for some task? You can find people to advise you on which headhunters are best at finding the right consultants for your circumstances.

In this ever-deepening tier of recursive explorations, there are also sidelong inquiries about the point of it all and about the chance that structural biases are built into our pyramid of knowledge. Several contributors write about the importance – and risks – of the democratization of knowledge being accomplished by the Internet. The movement to create Open Access to academic journals, for instance, promises to collapse the barriers of time, expense, and opportunity that have kept all but the elite few from acquaintance with the frontier in any field. But will the increasing transparency enabled – and demanded – by information technologies across all human endeavor be exploitable by parasites, like spam, music piracy, and their unnamed cousins waiting in the wings? Will our descendants learn how to be relatively immune to hype, as one contributor proposes, or will the spin doctors always stay a step or two ahead?

Cyberspace is still a largely anarchic territory, expanding faster than we can devise rules and principles to protect what needs protection. Will the arms race of yet further expansions of knowhow be at least a standoff, keeping society abreast of the problems? Will particular institutions prove especially vulnerable? Religions, for instance, have thrived for millennia in societies where knowledge could be severely restricted and are now discovering that they cannot erect a barrier able to hold back the flood of information, so they are searching for ways to preserve the allegiance of their members, by applying the resources of science and technology – surveys, focus groups, consultants, and every new media wrinkle. In the process, they are evolving faster than religions have ever evolved before. They are not alone in regretting the hyperavailability of information. What secret facts do you have a right *not* to know? Your IQ? Your chances of succumbing to Huntington's chorea or some other disease? Your popularity among your co-workers? On these and many other vexing questions we will need all the expert help we can get from those who think, for good reason, that they can see at least a little bit farther into the future than the rest of us. The seers assembled here are reassuringly hopeful.

But consider who they are, says the skeptic. They are all endowed with more than their fair share of position, security, and prestige. Not a one of them is poor, and a few are billionaires. No wonder they exude confidence and optimism! But even more important, I think, than security and prestige (and money), they are alike in being fortunate enough to be engaged on the cutting edge of whatever it is that matters the most to them. How many people in the world get to devote so much of their time and energy to such gratifying projects? Not enough – but, once again, knowledge is probably the key, and as we democratize access to it, more and more people will find the ways, and the time and energy, to construct meaningful lives, responsive to their own well-informed values. That will be a tumultuous world, but a better one.

Incredible Odds

MIHALY CSIKSZENTMIHALYI

Psychologist; director of the Quality of Life Research Center, Claremont Graduate University; author of *Flow: The Psychology of Optimal Experience*.

I am optimistic for the simple reason that given the incredible odds against the existence of entities that can ask such questions, of laptops on which to answer them, and so on – here we are, asking and answering!

Our Species Can Unravel Mysteries

BRIAN GREENE

Physicist, string theorist, Columbia University; author of *The Fabric of the Cosmos*.

As I help raise my two-year-old son, I witness a basic truth familiar to parents through the ages and across the continents – we begin life as uninhibited explorers with a boundless fascination for the ever growing world to which we have access. And what I find amazing is that if that fascination is fed, and if it's challenged, and if it's nurtured, it can grow to an intellect capable of grappling with such marvels as the quantum nature of reality, the energy locked inside the atom, the curved spacetime of the cosmos, the elementary constituents of matter, the genetic code underlying life, the neural circuitry responsible for consciousness, and perhaps even the very origin of the universe. While we evolved to survive, once we have the luxury of taking such survival for granted, the ability of our species to unravel mysteries grand and deep is awe-inspiring. I'm optimistic that the world will increasingly value the power of such rational thought and will increasingly rely on its insights in making the most critical decisions.

Good Choices Sometimes Prevail

JARED DIAMOND

Professor of geography, UCLA; author of Collapse: *How Societies Choose to Fail or Succeed.*

I am cautiously optimistic about the state of the world, because:

(1) Big businesses sometimes conclude that what is good for the long-term future of humanity is also good for their bottom line (note Wal-Mart's decision to shift its seafood purchases entirely to certified sustainable fisheries within the next three to five years).

(2) Voters in democracies sometimes make good choices and avoid bad choices (note last year's elections in a major first-world country).

The Decline of Violence

STEVEN PINKER

Psychologist, Harvard University; author of *The Blank Slate*.

In 16th-century Paris, a popular form of entertainment was cat burning, in which a cat was hoisted on a stage and slowly lowered into a fire. According to the historian Norman Davies, 'the spectators, including kings and queens, shrieked with laughter as the animals, howling with pain, were singed, roasted, and finally carbonized.'

As horrific as present-day events are, such sadism would now be unthinkable in most of the world. This is just one example of the most important and underappreciated trend in the history of our species: the decline of violence. Cruelty as popular entertainment, human sacrifice to indulge superstition, slavery as a labor-saving device, genocide for convenience, torture and mutilation as routine forms of punishment, execution for trivial crimes and misdemeanors, assassination as a means of political succession, pogroms as an outlet for frustration, and homicide as the major means of conflict resolution – all were unexceptional features of life for most of human history. Yet today they are statistically rare in the West, less common elsewhere than they used to be, and widely condemned when they do occur.

Most people, sickened by the headlines and the bloody history of the 20th century, find this claim incredible. Yet as far as I know, every systematic attempt to document the prevalence of violence over centuries and millennia (and, for that matter, over the past fifty years), particularly in the West, has shown that the overall trend is downward (though of course with many zigzags). The most thorough of such surveys is James Payne's *The History of Force*; other studies include Lawrence Keeley's *War Before Civilization,* Martin Daly and Margo Wilson's *Homicide*, Donald Horowitz's *The Deadly Ethnic Riot*, Robert Wright's *Nonzero*, Peter Singer's *The Expanding Circle*, Steven Leblanc's *Constant Battles,* and surveys of the ethnographic and archeological record by Bruce Knauft and Philip Walker.

Anyone who disputes this trend by pointing to residues of force in America (capital punishment in Texas, Abu Ghraib, sex slavery in immigrant groups, and so on) misses two key points. One is that statistically the prevalence of these practices is almost certainly a tiny fraction of what it was in centuries past. The other is that these practices are, to varying degrees, hidden, illegal, condemned, or at the very least (as in the case of capital punishment) intensely controversial. In the past, they were no big deal. Even the mass murders of the 20th century in Europe, China, and the Soviet Union probably killed a smaller proportion of the population than a typical biblical conquest or hunter-gatherer feud. The world's population has exploded, and wars and killings are scrutinized and documented, so we are more aware of violence even though it may be statistically less extensive.

What went right? No one knows, possibly because we have

been asking the wrong question: 'Why is there war?' instead of 'Why is there peace?' There have been some suggestions, all unproved. Perhaps the gradual perfecting of a democratic Leviathan – 'a common power to keep [us] in awe' – has removed the incentive to do it to them before they do it to us. Payne suggests that it's because, for many people, life has become longer and less awful; when pain, tragedy, and early death are expected features of one's own life, one feels fewer compunctions about inflicting them on others. Wright points to technologies that enhance networks of reciprocity and trade, which make other people more valuable alive than dead. Singer attributes it to the inexorable logic of the golden rule: The more one knows and thinks, the harder it is to privilege one's own interests over those of other sentient beings. Perhaps this is amplified by cosmopolitanism, in which history, journalism, memoir, and realistic fiction make the inner lives of other people, and the contingent nature of one's own station, more palpable – the feeling that 'There but for fortune go I.'

My optimism lies in the hope that the decline of violence over the centuries is a real phenomenon, that it is the product of systematic forces that will continue to operate, and that we can identify those forces and perhaps concentrate and bottle them.

War Will End

JOHN HORGAN

Director of the Center for Science Writings, Stevens Institute of
Technology; author of *Rational Mysticism*.

I'm optimistic that one day war – large-scale, organized, group
violence – will end once and for all.

Many people find my optimism naïve, if not delusional.
Recently I taught a class called 'War and Human Nature' and
my students polled their classmates on the following question:
'Do you think humanity will ever stop fighting wars once and
for all time?' Of the 205 respondents, 185 replied 'No' and
only 20 said 'Yes' or 'Maybe.' Several of the 'optimists' added
comments like, 'Yes, war will end when the human race will
end' and 'Yes, because in the future the human species will
unite to fight alien species.'

Recent scholarship on warfare seems at first glance to support
this fatalism. Just a few decades ago, many scholars believed in
the myth of the peaceful savage, which depicts war as a by-prod-
uct of modern civilization that did not exist in pre-state
societies. In his book *Constant Battles*, the anthropologist Steven
LeBlanc debunks this myth, pointing out that the vast majority
of primitive, pre-state societies engaged in at least occasional
warfare. Mortality rates from violence in some societies reached
as high as 50 percent.

But these grim statistics yield a surprisingly upbeat message:
Things are getting better! Hard as it may be to believe, humanity
has become much less violent than it used to be. In fact civilization, far from creating the problem of warfare, is apparently

helping us to solve it. In *War Before Civilization*, the anthropologist Lawrence Keeley estimates that in the blood-soaked 20th century 100 million men, women, and children died from war-related causes, including disease and famine. The total would have been 2 billion, Keeley notes, if our rates of violence had been as high as in the average primitive society. Moreover, as the scholars Charles Kurzman and Neil Englehart point out in their 2006 essay 'Welcome to World Peace,' conventional wars between the armies of two or more nations, and even civil wars, have become much less common in recent decades. We are now dealing primarily with guerrilla wars, insurgencies, terrorism – or what the political scientist John Mueller calls 'the remnants of war.'

These statistics do not provide much solace to the victims of war's remnants in Iraq, Darfur, Sri Lanka, Palestine, Colombia, and other troubled regions, but they show that we are moving in the right direction. Other recent events offer more grounds for optimism. As recently as the late 1980s, we faced the threat of a global nuclear holocaust. Then, incredibly, the Soviet Union dissolved and the cold war ended peacefully. South Africa's apartheid also ended, without significant violence, and human rights have advanced elsewhere around the world.

The first and crucial step toward ending war is to believe we can do it. We should also recognize that war is overdetermined – stemming from many different possible causes – and so peace must be overdetermined, too. In their final papers, most of my students wisely advocated pursuing not a single, silver-bullet solution to the problem of war but multiple approaches. Their proposals included supporting democracy in other countries, bolstering the U.N.'s peacekeeping efforts, fighting poverty and improving education, restricting or eliminating arms sales, inculcating tolerance for other cultures in children, and giving women more of a role in government. 'Achieving peace on a global level will not be easy,' one student wrote, 'but things already seem to be moving in the right direction. Humanity's best shot at ending war is now.'

His optimism fuels my optimism.

World Peace

JOHN McCARTHY

Computer scientist; first-generation artificial-intelligence pioneer, Stanford University.

I'm optimistic about the sustainability of material progress, but since I'm known for that, I'll refrain. Instead I want to express optimism about world politics, especially about world peace.

World peace is what we have. There are only minor wars and no present prospect of a major war threatening Western civilization and its present extensions to the developing countries. Only Africa and the Arab world are in bad shape. Contrast this with the time between 1914 and 1989, when there were serious attempts at world domination accompanied by at least three genocides.

Admittedly something bad and surprising could happen. One hundred years ago, in 1907, no one predicted such troubles as happened. Even in April 1914, Bertrand Russell could write: 'To us, to whom safety has become monotony, to whom the primeval savageries of nature are so remote as to become a mere pleasing condiment to our ordered routine, the world of dreams is very different from what it was amid the wars of Guelf and Ghibelline.'

As for Arab jihadism, I think they'll get over it as soon as a new generation matures to oppose their parents' slogans. If not:

> Whatever happens we have got
> The Maxim Gun, and they have not.
> – Hilaire Belloc, *The Modern Traveller*

It is important that the political causes of the 20th-century disasters, virulent and militaristic nationalism accompanied by letting one man take power, do not exist in major countries today. Communism is dead as a motivator of violence. The green movement is accompanied by occasional minor violence, but a green Hitler or Stalin seems unlikely.

Still, it's hard to predict a hundred years ahead. As Stephen Hawking advocates, humanity would be safer if it expanded beyond the earth.

We Are Making Moral Progress

SAM HARRIS

Neuroscience researcher; author of *Letter to a Christian Nation.*

No one has ever mistaken me for an optimist. Yet when I consider what is perhaps the most pristine source of pessimism – the moral development of our species – I find reasons for hope.

Despite our perennial mischief, I believe we have made unmistakable progress in our morality. Our powers of empathy appear to be growing. We seem to be more likely now than at any point in our history to act for the benefit of humanity as a whole.

Of course, the 20th century delivered some unprecedented horrors. But those of us living in the developed world are becoming increasingly alarmed by our capacity for doing one another harm. We are less tolerant of 'collateral damage' in war – undoubtedly because we now see images of it – and we are less comfortable with ideologies that demonize whole groups of human beings, justifying their abuse or outright destruction.

Taking a somewhat provincial example: Racism in the United States has unquestionably diminished. If you doubt this, consider the following *Los Angeles Times* editorial, written in 1910, in response to Jack Johnson's successful heavyweight title defense against Jim Jeffries, the so-called 'Great White Hope':

> A Word to the Black Man:
>
> Do not point your nose too high.
> Do not swell your chest too much.
> Do not boast too loudly.
> Do not be puffed up.
> Let not your ambition be inordinate
> Or take a wrong direction.
> Remember, you have done nothing at all.
> You are just the same member of society
>   you were last week.

You are on no higher plane,
Deserve no new consideration,
And will get none.
No man will think a bit higher of you
Because your complexion is the same
As that of the victor at Reno.

A modern reader could be forgiven for thinking that this dollop of racist hatred was put out by the Ku Klux Klan. Rather, it represented the measured opinion of one of the most prominent newspapers in the United States. Is it conceivable that our mainstream media will once again give voice to such racism? I think it far more likely that we will proceed along our current path: Racism will continue to lose its subscribers; the history of slavery in the United States will become even more flabbergasting to contemplate; and future generations will marvel at the ways we, too, failed in our commitment to the common good. We will embarrass our descendants, just as our ancestors embarrass us. This is moral progress.

I am bolstered in my optimism by the belief that morality is a genuine sphere of human inquiry, not a mere product of culture. Morality, rightly construed, relates to questions of human and animal suffering. This is why we don't have moral obligations toward inanimate objects (and why we *will* have such obligations toward conscious computers, if we ever invent them). To ask whether a given action is right or wrong is really to ask whether it will tend to create greater well-being or greater suffering for oneself and others. And there seems little doubt that there are right and wrong answers here. This is not to say that there will always be a single right answer to every moral question, but there will be a range of appropriate answers, as well as answers that are clearly wrong. Asking whether or not an action is good or bad may be like asking whether a given substance is healthy or unhealthy to eat: There are of course many foods that are appropriate to eat, but there is also a biologically important (and objective) distinction between food and poison.

I believe that there are right and wrong answers to moral questions in the same way that there are right and wrong answers to questions about biology. This commits me to what philosophers often call moral realism – as opposed to anti-realism, pragmatism, relativism, postmodernism, or any other view that places morality entirely in the eye of the beholder. It is often thought that moral realism fails because it requires that moral truths exist independent of minds (it doesn't). Indeed, this worry partly explains humanity's enduring attachment to religion, for many people believe that unless we keep our moral intuitions pegged to the gold standard of God's law, we cannot say that anyone is ever right or wrong in objective terms.

Consider the phenomenon of 'honor killing.' Throughout much of the Muslim world at this moment, women are thought to dishonor their families by refusing to enter into an arranged marriage, seeking a divorce, committing adultery – or even by getting raped. Women in these situations are often murdered by their fathers, husbands, or brothers, sometimes with the collaboration of other women. Is honor killing wrong? I have no doubt that it is. But is it *really* wrong?

There seems to be no question that we are wired in such a way that love is more conducive to happiness than are hate, fear, and shame. If this is true, honor killing would be wrong even if a majority of human beings agreed that it was right. It would be wrong because this practice (along with the intentions that give rise to it) reliably diminishes human happiness: It creates immense suffering for women and girls; it conditions men to feel that their personal dignity is predicated on something it need not be predicated on; it deranges the relationships between men and women, making them far less loving and compassionate (and therefore a lesser source of happiness) than they might otherwise be. While these are claims about human subjectivity, they are also, at bottom, *objective* claims about the real foundations of human happiness.

All of this implies, of course, that morality is a potential branch of scientific inquiry – not merely that science will one day describe our moral judgments at the level of the brain but that science may one day be able to tell us what is good (that is, it will tell us which psychological intentions and social practices are truly conducive to the deepest happiness).

Because I believe that moral truths transcend the contingencies of culture, I think that human beings will eventually converge in their moral judgments. I am painfully aware, however, that we live in a world where Muslims riot by the hundreds of thousands over cartoons, where Catholics oppose condom use in villages decimated by AIDS, and where the only 'moral' judgment that seems guaranteed to unite the better part of humanity at this moment is that homosexuality is wrong. Which is to say that I am here celebrating our moral progress while being convinced that billions of my neighbors are profoundly confused about good and evil. I may be a bigger optimist than I thought.

The Unending Stream of Bad News Is Itself Flawed

CHRIS ANDERSON

Curator of the annual TED (Technology, Entertainment, and Design) Conference.

Paradoxically, one of the biggest reasons for being optimistic is that there are systemic flaws in the reported worldview. Certain types of news (for example, dramatic disasters and terrorist actions) are significantly overreported, others (such as scientific progress and meaningful statistical surveys of the state of the world) are significantly underreported. Although this imbalance leads to major problems – such as distortion of rational public policy and a perpetual gnawing fear of apocalypse – it is also reason to be optimistic. Once you realize you're being brainwashed to believe that things are worse than they are, you can, with a little courage, step out into the sunshine.

How does the deception take place?

The problem starts with a deep human psychological response. We're wired to react more strongly to dramatic stories than to abstract facts. One can readily imagine possible historical and Darwinian reasons why this might be so. The news that an invader has just set fire to a hut in your village demands immediate response. Genes promoting equanimity in such circumstances would have burned up long ago.

Although our village is now global, we still instinctively react the same way. Spectacle, death, and gore. We lap it up. Layer on top of that a media economy driven by competition for attention, and the problem is magnified. Over the years, media owners have proved to their complete satisfaction that the stories that attract large audiences are the simple human dramas. 'ROTTWEILER SAVAGES BABY' is a bigger story than 'POVERTY PERCENTAGE FALLS,' even though the latter is a story about better lives for millions.

Today our media can source news from 190 countries and 6 billion people. Therefore you can be certain that every single day there will be word of spectacularly horrifying things happening somewhere. And should you get bored of reading about bombs, fires, and wars, why not see them breaking live on cable TV twenty-four/seven, with ever more intimate pictures and emotional responses. Meta-level reporting doesn't get much of a look-in.

So, for example, the publication of a report last year by the University of British Columbia's Human Security Center received little attention, despite the fact that it concluded that the numbers of armed conflicts in the world had fallen 40 percent in little over a decade and that the number of fatalities per conflict had also fallen. Think about that. The entire news agenda for a decade, received as endless tales of wars, massacres, and bombings, actually missed the key point. Things are getting better. If you believe Robert Wright's optimistic view of human history as articulated in his *Nonzero: The Logic of Human Destiny* (2000), this improvement is part of a long-term (and admittedly volatile) trend in which cooperation eventually trumps conflict. The percentage of males estimated to have died in violence in hunter-gatherer societies? Approximately 30 percent. The percentage of males who died violently in the 20th century complete with two world wars and a couple of nukes? Approximately 1 percent. Trends for violent deaths so far in the 21st century? Falling. Sharply.

In fact, most meta-level reporting of trends shows a world that is getting better. We live longer, in cleaner environments; we are healthier; we have access to goods and experiences that kings of old could never have dreamed of. If that doesn't make us happier, we have no one to blame but ourselves – oh, and the media lackeys who continue to feed us the woeful litany we subconsciously crave.

Techno-Optimism and the Energy Challenge

MARTIN REES (LORD REES OF LUDLOW)

President of the Royal Society; professor of cosmology and astrophysics; master of Trinity College, Cambridge University; author of *Our Final Century?*

A few years ago, I wrote a short book entitled *Our Final Century?* I guessed that, taking all risks into account, there was only a 50-percent chance that civilization would get through to 2100 without a disastrous setback – a far from cheerful conclusion. However, I was surprised by the way my colleagues reacted to the book. Many thought a catastrophe was even more likely than I did and regarded me as an optimist. I stand by this optimism.

There are indeed powerful grounds for being a techno-optimist. For most people in most nations, there's never been a better time to be alive. The innovations that drive economic advance – information technology, biotech, and nanotech – can boost the developing as well as the developed world. We're becoming embedded in a cyberspace that can link anyone, anywhere, to all the world's information and culture – and to every other person on the planet. Creativity in science and the arts is open to hugely more people than in the past. Twenty-first-century technologies will offer lifestyles that are environmentally benign – involving lower demands on energy or resources than what we'd consider constituted a good life today. And we could readily raise the funds – were there the political will – to lift the world's 2 billion most deprived people from their extreme poverty.

Later in this century, mind-enhancing drugs, genetics, and 'cyborg' techniques may change human beings themselves. That's something qualitatively new in recorded history – and it will pose novel ethical conundrums. Our species could be transformed and diversified (here on Earth and perhaps beyond) within just a few centuries.

The benefits of earlier technology weren't achieved without taking risks – we owe modern aviation and modern surgery to many martyrs. But though plane crashes, boiler explosions, and the like were horrible, there was a limit to just how horrible – a limit to their scale. In our increasingly interconnected world, where technology empowers us more than ever, we're vulnerable to scary new risks – events of such catastrophic global consequences that it's imprudent to ignore them even if their probability seems low.

One set of risks stems from humanity's collective impact. Our actions are transforming, even ravaging, the entire biosphere – perhaps irreversibly – through global warming and loss of biodiversity. Remedial action may come too late to prevent runaway climatic or environmental devastation. We also face vulnerabilities of a quite different kind, stemming from unintended consequences (or intended misuse) of ever more empowering bio- and cybertechnology. The global village will have its village idiots.

The risks are real, but by making the right collective choices we can alleviate all these hazards.

Among such choices, my number-one priority would be much-expanded research and development into a whole raft of techniques for storing energy and generating it by 'clean,' or low-carbon, methods. The stakes are high – the world spends nearly $3 trillion per year on energy and its infrastructure. This effort can engage not just those in privileged technical environments in advanced countries but a far wider talent pool. Even if we discount climate change completely, the quest for clean energy is worthwhile on grounds of energy security, diversity, and efficiency.

This goal deserve a priority and commitment from governments akin to that accorded to the Manhattan Project or the Apollo moon landing. It should appeal to the idealistic young – indeed, I can't think of anything that would do more to attract the brightest and best of them into science than a strongly proclaimed commitment, from all technologically developed nations, to take a lead in providing clean and sustainable energy for the developing and the developed world.

The Divide Between Scientific Thinking and the Rest of Our Culture Is Decreasing

CARLO ROVELLI

Physicist, Université de la Mediterranée, Marseille; author of *What Is Time? What Is Space?*

Some days I wake up optimistic, others not at all. When I am optimistic, I think that humans are increasingly realizing that rational thinking is better for them than irrational thinking. In the process, scientific thinking is growing in depth, abandoning a certain traditional superficiality, regaining contact with the rest of the culture, learning to deal with the full complexity of the human experience and the search for knowledge. Nonscientific thinking is still everywhere, but it is losing ground.

In the small world of academia, the senseless divide between science and the humanities is slowly evaporating. Intellectuals on both sides realize that the complexity of contemporary knowledge cannot be fully grasped unless we look at it all. A contemporary philosopher who ignores scientific thinking is disconnected from reality, but an increasing number of theoretical physicists are also realizing that to solve quantum gravity (for instance), we cannot avoid addressing foundational 'philosophical' questions. And more and more scientists are coming out of the lab and speaking out.

When I am pessimistic, I think that history shows that human madness is with us to stay: war, greed for more power and more riches, religion as the depository of ultimate truth, fear of those different from us. I see all this madness solidly in control of the planet's affairs.

When I am optimistic, I think that the past was worse: We are definitely moving toward a better and more reasonable world. There are countries today that have not started a war in decades, and in fact these countries are the majority. This is something new in the history of the world. The number of people who have realized how much of religious belief is non-sensical continues to grow, and no doubt this will help decrease belligerency and intolerance.

There is progress on the religious front, too. In a recent interview on CNN, the Dalai Lama was asked how it feels to be the leader of a major religion in a secular world. He smiled and answered that he was happy to see that the modern world had a rich *secular* spiritual life. A secular spiritual life, he explained, is a life that is rich intellectually as well as emotionally. The next question was whether he really believed he was the Dalai Lama, the reincarnation of previous Dalai Lamas. This time he laughed and answered, 'Of course I am the Dalai Lama.' But to be the reincarnation of previous lamas, he continued, did not mean to 'be them,' it meant to continue something they had been developing. Not all our major religious leaders are so reasonable. But if one of them can be, can we not hope, at least in our optimistic moments, that others will follow?

Twenty-six centuries have lapsed since Anaximander suggested that rain is not sent by Zeus but is water evaporated by the sun and carried by the wind. The battle to realize that the scientific method of representing knowledge and the science-minded mode of thinking is deeper, richer, and better for us than any God is still ongoing, but by no means is it lost, as it often seems.

The Evaporation of the Powerful Mystique of Religion

DANIEL C. DENNETT

Philosopher; university professor, codirector of the Center for Cognitive Studies, Tufts University; author of *Breaking the Spell: Religion as a Natural Phenomenon*.

I'm so optimistic that I expect to live to see the evaporation of the powerful mystique of religion. In about twenty-five years, almost all religions will have evolved into very different phenomena, so much so that in most quarters religion will no longer command the awe it does today. Of course, many people – perhaps a majority of people in the world – will still cling to their religion with the sort of passion that can fuel violence and other intolerant and reprehensible behavior. But the rest of the world will see this behavior for what it is and learn to work around it until it subsides, as it surely will. That's the good news.

The bad news is that we will need every morsel of this reasonable attitude to deal in an effective way with such complex global problems as climate change, increasing scarcity of fresh water, and economic inequality. It will be touch and go, and in my pessimistic mood I think Sir Martin Rees may be right: Some disaffected religious (or political) group may unleash a biological or nuclear catastrophe that forecloses all our good efforts. But I do think we have the resources and the knowledge to forestall such calamities, if we are vigilant.

Recall that only fifty years ago smoking was a high-status activity, and it was considered rude to ask somebody to stop smoking in one's presence. Today we've learned that we shouldn't make the mistake of trying to prohibit smoking altogether, and so we still have plenty of cigarettes and smokers, but we have certainly contained the noxious aspects within acceptable boundaries. Smoking is no longer cool, and the day will come when religion is, first, a take-it-or-leave-it choice and later no longer cool — except in its socially valuable forms, where it will be one type of allegiance among many. Will those descendant institutions still be religions, or will religions have thereby morphed themselves into extinction? It all depends on what you think the defining elements of religion are. Are dinosaurs extinct, or do their lineages live on as birds?

Why am I confident that this metamorphosis of religion will happen? Mainly because of the asymmetry in the information explosion. With the worldwide spread of information technology (not just the Internet but cell phones and portable radios and television), it is no longer feasible for guardians of religious traditions to protect their young from exposure to the kinds of facts (and, yes, of course, misinformation and junk of every genre) that gently, irresistibly undermine the mindsets requisite for religious fanaticism and intolerance. The religious fervor of today is a last, desperate attempt by our generation to block the eyes and ears of the coming generations, and it isn't working. For every well-publicized victory — the inundation of the Bush administration with evangelicals, the nation's growing number of homeschoolers, the rise of radical Islam, the much exaggerated 'rebound' of religion in Russia following the collapse of the Soviet Union, to take the most obvious cases — there are many less dramatic defeats, as young people quietly walk away from the faith of their parents and grandparents. That trend will continue, especially when young people come to know how

many of their peers are making this low-profile choic/
the world, the category of 'not religious' is growing fas.
the Mormons, faster than the evangelicals, faster even than
Islam, whose growth is due almost entirely to fecundity, not
conversion, and is bound to level off soon.

Those who are secular can encourage their children to drink
from the well of knowledge wherever it leads them, confident
that only a small percentage will rebel against their secular
upbringing and turn to one religion or another. Cults will rise
and fall, as they do today and have done for millennia, but
only those that can metamorphose into socially benign organizations will flourish. Many religions have already made the
transition, quietly de-emphasizing the irrational elements in
their heritages, abandoning the xenophobic and sexist prohibi-
tions of their recent past, and turning their attention from
doctrinal purity to moral effectiveness. The fact that these
adapting religions are scorned as *former* religions by the diehard
purists shows how brittle the objects of their desperate allegiance have become. As the world informs itself about these
transitions, those who are devout in the old-fashioned way will
have to work around the clock to provide attractions, distractions – and guilt trips – to hold the attention and allegiance of
their children. They will not succeed, and it will not be a painless transition. Families will be torn apart, and generations will
accuse each other of disloyalty and worse: The young will be
appalled by their discovery of the deliberate misrepresentations
of their elders, and their elders will feel abandoned and betrayed
by their descendants. We must not underestimate the anguish
that these cultural transformations will engender, and we should
try to anticipate the main effects and be ready to provide relief
and hope for those who are afflicted.

I think the main problem we face today is overreaction, making martyrs out of people who desperately want to become martyrs. What it will take is patience, good information, and a steady demand for universal education about the world's religions. This will favor the evolution of avirulent forms of religion, which we can all welcome as continuing parts of our planet's cultural heritage. Eventually the truth will set us free.

A Proper Scientific Understanding of Irrationality (and Religion in Particular)

ANDREW BROWN

Journalist, *The Guardian*; author of *The Darwin Wars*.

I'm not actually optimistic about anything very much, but it's clear that if civilization is to survive we need a proper scientific understanding of irrationality in general and religion in particular. To be optimistic about that is a precondition for optimism about anything else. What might such an understanding look like?

To begin with, it would be naturalistic and empirical. It would not start from definitions of religion or faith but from a careful study, in the spirit of William James, of how religious people actually behave and believe. What would be found, again in a Jamesian spirit, is that there are varieties of religious behavior, as there are varieties of religious experience. We would need to know how these experiences are related to one another and to those experiences not described as religious. It may well be that 'religion' is a concept no more useful than 'phlogiston.'

It would take seriously Dan Dennett's distinction between beliefs and opinions – more seriously, I think, than he sometimes does himself. A belief, in Dennett's sense, is a kind of propensity to behave as if certain things were true. It need not be conscious. The kind of conscious, articulable propositions about the world which most people mean by 'belief' he calls 'opinion.'

In this sense, an inquiry into religious belief would be distinct from an inquiry into religious opinions: Religious 'belief' would involve the largely unconscious mechanisms that lead people to behave superstitiously or reverently or with a disdain for heretics; religious opinions would be the reasons they give for this behavior. We need to understand both. It may be that the opinions of the religious correspond to their beliefs, but that is something to be established in every case by empirical inquiry. It's obvious that in most cases they don't. Intellectuals are supposed to be motivated by their opinions; some of them actually are. But all of us are motivated by our beliefs and prejudices.

In particular, such an inquiry would be careful about what counts as evidence. A friend of mine who does consciousness research once said sourly that 'the problem with the brain is that if you go looking for something in there, you're very liable to find it.' Similarly, if you go looking for some particular quality in religious belief, you are likely to find it there. The problem is that you are very likely to find its opposite as well. What's needed to solve the problem is the distinctly scientific attitude that takes disconfirming evidence seriously and doesn't respond to it by simply repeating the confirming evidence.

That is lacking in most atheist propaganda: Even the stuff that comes from scientists – especially the stuff that comes from some scientists – completely ignores disconfirming evidence about religion.

Last winter I saw *On Religion*, a play by the British atheist philosopher A. C. Grayling, which is an excellent dramatization of some of these issues. The atheist character, a woman lecturer, recounts the story of a scientist who has spent fifteen years arguing that the cell's Golgi apparatus does not exist and is an artifact of the inadequacies of our microscopes. Finally, he attends a lecture from a visiting cell biologist, who proves conclusively that the Golgi apparatus does in fact exist. Just as the whole department is trying to avoid his eye in sympathetic shame, he rushes up to the lecturer, grabs his hand, and says, 'My dear fellow, I wish to thank you. I have been wrong these fifteen years.' It is an improving and inspiring story, which pitches over into bathos as soon as the atheist spells out the moral. 'No religious person could ever say that,' she says. Has Grayling really never heard of the phenomenon of conversion? What do the converted say, if not that some evidence has convinced them that they were wrong all their lives before?

So, I think – if I am to be optimistic – that there will be a breakthrough in the empirical study of religion, at the end of which no scientist will ever feel able to assert that 'no religious person could ever say . . . ' without having made a careful inquiry into what religious people actually do say and what they mean by it.

The Final Scientific Enlightenment

RICHARD DAWKINS

Evolutionary biologist and Charles Simonyi Professor for the Understanding of Science, University of Oxford; author of *The God Delusion*.

I am optimistic that the physicists of our species will complete Einstein's dream and discover the final theory of everything before superior creatures, evolved on another world, make contact and tell us the answer. I am optimistic that, although the theory of everything will bring fundamental physics to a convincing closure, the enterprise of physics itself will continue to flourish, just as biology went on growing after Darwin solved its deep problem. I am optimistic that the two theories together will furnish a totally satisfying naturalistic explanation for the existence of the universe and everything that's in it, including ourselves. And I am optimistic that this final scientific enlightenment will deal an overdue deathblow to religion and other juvenile superstitions.

Science and the Decline of Magic

MICHAEL SHERMER

Publisher of *Skeptic* magazine, monthly columnist for *Scientific American*; author of *Why Darwin Matters*.

I am optimistic that science is winning out over magic and superstition. That may seem irrational, given the data from pollsters on what people believe. For example, a 2005 Pew Research Center poll found that 42 percent of Americans believe that 'living things have existed in their present form since the beginning of time.' The situation is even worse when we examine other superstitions, such as these percentages of belief, published in a 2002 National Science Foundation study:

ESP	60%
UFOs	30%
Astrology	40%
Lucky numbers	32%
Magnetic therapy	70%
Alternative medicine	88%

Nevertheless, I take the historian's long view: Compared to what people believed before the scientific revolution of the 17th century, there is much cause for optimism. Consider what people believed a mere four centuries ago, just as science began lighting candles in the dark. In 16th- and 17th-century England, for example, almost everyone believed in sorcery, werewolves, hobgoblins, witchcraft, astrology, black magic,

demons, prayer, and providence. 'A great many of
be in trouble, or sickness, or lose anything, we ru
thither to witches, or sorcerers, whom we call
seeking aid and comfort at their hands,' noted B
in 1552. Saints were worshiped. Liturgical books provided rit-
uals for blessing cattle, crops, houses, tools, ships, wells, and
kilns, not to mention sterile animals, infertile couples, and the
sick. In his 1621 book *Anatomy of Melancholy*, Robert Burton
explained: 'Sorcerers are too common; cunning men, wizards,
and white witches, as they call them, in every village, which, if
they be sought unto, will help almost all infirmities of body and
mind.'

Just as alcohol and tobacco were essential anesthetics for the
easing of pain and discomfort, superstition and magic were the
basis for the mitigation of misfortune. As the great Oxford historian of the period, Keith Thomas, writes in his classic 1971
work *Religion and the Decline of Magic:* 'No one denied the
influence of the heavens upon the weather or disputed the rel-
evance of astrology to medicine or agriculture. Before the 17th
century, total skepticism about astrological doctrine was highly
exceptional, whether in England or elsewhere.' And it wasn't
just astrology. 'Religion, astrology and magic all purported to
help men with their daily problems by teaching them how to
avoid misfortune and how to account for it when it struck.'
With such sweeping power over nearly everyone, Thomas concludes, 'If magic is to be defined as the employment of
ineffective techniques to allay anxiety when effectives ones are
not available, then we must recognize that no society will ever
be free from it.' The superstitious we will always have with us.

Nevertheless, the rise of science ineluctably attenuated this
near universality of magical thinking by proffering natural

planations where before there were only supernatural ones. Before Darwin, design theory (in the form of William Paley's natural theology, which gave us the 'watchmaker' argument) was the only game in town, so everyone believed that life was designed by God. Today fewer than half of all Americans believe that (in the most religious nation of the developed democracies), and in most other parts of the world virtually everyone accepts evolution without qualification. That's progress.

The rise of science even led to a struggle to find evidence for superstitious beliefs that previously needed no propping up with facts. Consider the following comment from an early 17th-century book that shows how, even then, savvy observers grasped the full implications of denying the supernatural altogether: 'Atheists abound in these days and witchcraft is called into question. If neither possession nor witchcraft (contrary to what has been so long generally and confidently affirmed), why should we think that there are devils? If no devils, no God.'

Magic transitioned into empirical magic and formalized methods of ascertaining causality by connecting events in nature – the very basis of science. As science grew in importance, the analysis of portents was often done meticulously and quantitatively, albeit for purposes both natural and supernatural. As one diarist privately opined in 1682 on the nature and meaning of comets: 'I am not ignorant that such meteors proceed from natural causes, yet are frequently also the presages of imminent calamities.'

Science arose out of magic, which it ultimately displaced. By the 18th century, astronomy had replaced astrology, chemistry had succeeded alchemy, probability theory had dislodged belief in luck and fortune, city planning and social hygiene had ameliorated disease, and the grim vagaries of life became less grim, and less vague. As Francis Bacon concluded in his 1626 work *New Atlantis*: 'The end of our foundation is the knowledge of causes and the secret motions of things and the enlarging of the bounds of human empire, to the effecting of all things possible.'

Sic itur ad astra: Thus do we reach the stars.

Reliance on Evidence

CLAY SHIRKY

Social and technology network topology researcher; adjunct professor, Interactive Telecommunications Program, NYU's Tisch School of the Arts.

As schoolchildren, we learn that different weights fall at the same speed. This simple and readily tested observation, first published by Galileo, refuted Aristotle, who claimed that heavy things fall faster. As Galileo put it in *Two New Sciences,* 'I greatly doubt that Aristotle ever tested by experiment whether it be true . . .' We are left to wonder how people simply believed what they were told, and for two millennia at that, without ever checking. Surely the power of evidence over authority is obvious.

Except it isn't. Even today, evidence has barely begun to upend authority; the world is still more in thrall to Aristotle than to Galileo. To give one homely example, the time-honored advice for those suffering from bad backs has been bed rest; only recently have we discovered that bed rest isn't the best treatment and isn't even particularly good compared to moderate activity. How did this discovery come about? A researcher in the field of evidence-based medicine surveyed multiple databases of trials and results for patients with back pain. (It tells us something about medicine's current form that we even need a term like 'evidence-based medicine.') And why did it take so long to look at the evidence? For the same reason it took so long to question Aristotle: Some doctor in the distant past reasoned that bed rest would be a good idea, and once it became the authoritative view no one thought to question it.

In school, the embrace of evidence is often taught as if it were a one-time revolution that has long since been internalized by society. It hasn't. The idea of evidence is consistently radical: Take nothing on faith; no authority is infallible; if you figure out a good way to test something, you can contradict hallowed figures with impunity.

Evidence will continue to improve society, but slowly – this is long-view optimism. The use of evidence dragged the curious mind from the confusion of alchemy into the precision of chemistry in the historical blink of an eye, but its progress beyond the hard sciences has been considerably slower. Even accepting the notion that evidence ought to shape our views is inconsistent with much human behavior. Everything from the belief in supernatural beings to deference to elders militates against the idea that a single person, if he or she comes to understand the evidence, should be allowed to upend a millennium of cherished belief.

It is only in the last hundred years that evidence has begun spreading from the hard sciences into other parts of human life. Previous platitudes about the unpredictability or universal plasticity of human behavior are giving way to areas of inquiry with names like sociobiology, evolutionary psychology, and behavioral economics. (That we need labels like 'behavioral economics' and 'evolutionary psychology' says as much about those fields as 'evidence-based medicine' does about medicine.) As reliance on evidence spreads, it takes with it an understanding of how evidence works. Apologists for religion often bolster their claims by noting that one cannot disprove the existence of supernatural beings. This argument assumes that their listeners don't understand how evidence works: It makes sense to believe in things for which there is evidence and no sense to believe in things for which there is none. As evidence moves out of the lab and into everywhere else, rhetorical tricks like that are going to be progressively less effective. There will still be fundamentalists, of course – probably more fervent ones, as improved evidence requires a heightened ability to shield the mind – but the oxymoronic middle ground of 'religious but reasonable' will become progressively harder to occupy.

This isn't just about religion, though. Most of the really important parts of our lives – whom we love and how, how we live and why, why we lie and when – have yet to yield their secrets to real evidence. We will see a gradual spread of things like evidence-based politics and law: What is the evidence that this expenditure, or that proposed bill, will have the predicted result? The expectation that evidence can answer questions about the structure of society will discomfit every form of government that relies on sacrosanct beliefs. Theocracy and communism are different in many ways, but they share the same central weakness: They are based on some set of assertions that must remain beyond question.

Social science is expanding because we are better about gathering data and understanding them. We have gone from a drought to a flood of data about personal and social behavior in the last generation. We will learn more about the human condition in the next two decades than we did in the last two millennia, and we will then begin to apply what we learn, everywhere. Evidence-based treaties. Evidence-based teaching. Evidence-based design. Evidence-based parenting.

There will always be some questions we can't answer, but they will be closer in spirit to 'Who put the *bomp* in the *bomp-bah-bomp-bah-bomp?*' than to 'Why do fools fall in love?' There is an astonishing amount of work going on on that latter question right now, and there's a reasonable chance we'll have a really good answer – to it and to thousands of other questions once thought to be beyond study or explanation – in the coming years.

Evidence-Based Decision Making Will Transform Society

J. CRAIG VENTER

Decoder of the human genome, director of the J. Craig Venter Institute.

I am optimistic that one of the key tenets of scientific investigation, evidenced-based decision making, will be extended to all aspects of modern society. Good experimental design works toward creating conditions that provide the most useful information on a given topic while attempting to eliminate, or at least limit, spurious, irrelevant artifacts that could falsely influence data interpretation. Data are collected until a threshold is exceeded permitting either conclusions to be drawn or the development of a hypothesis that, with further testing, can be validated or falsified.

Not all questions can be answered just by looking at the evidence, because we are still at a very early stage in understanding our universe. For example, in trying to understand how life began on our planet, we can only guess, based on certain assumptions, whether it originated *de novo* here or arrived from another planet or a distant galaxy. We do know that a few hundred kilograms of material are exchanged annually between Earth and Mars and that new planets are being discovered at an unprecedented pace. When we discover microbial life on Mars, we will double the number of planets with known life while increasing the possibility of finding life elsewhere in the universe.

For most scientists, the evidence for evolution, regardless of its origins, has been overwhelming. The fossil record was sufficient evidence for most, but now, with genome-sequencing information from all branches of life including some of our closest relatives – like Neanderthals, chimps, and rhesus monkeys – the results should be clear-cut for anyone whose thinking is not overly clouded by a belief system.

In contrast, we have newspapers and radio and television news stations presenting subjective, selective subsets of information and political campaigns and pronouncements that can only be dismissed as partisan. We need to push harder for an education system that teaches evidence-based decision making – and we need to hold our public leaders to a higher standard and less partisan behavior – as we tackle some of the most difficult challenges facing the future of humanity.

Human Beings Are Different from Their Ancestral Species

DOUGLAS RUSHKOFF

Media analyst; documentary writer; author of *Get Back in the Box:
Innovation from the Inside Out.*

Now that we've gotten false notions of 'God' out of the way,
we come up against the question from which he insulated
us: If human beings are not the 'chosen species,' then are we
at least capable of transcending nature, from which we
emerged?

Our most natural inclination should be to kill each other,
one way or another. From plankton to pachyderms, the myth of
nature as a sustainable and loving collaborative is about as
absurd as that of a Creator. Unless we prove different from
every other species, we will continue to compete with the rest of
the planet's inhabitants for a disproportionate share of its
resources – and with one another for the spoils of this ongoing
war. That's just life.

I'm optimistic that human beings can be different from our
ancestral species and that the endless comparisons between
human behavior and that of other species are, ultimately, misleading. I hope that although sponge colonies fight endlessly
with those of a different color, this need not mean that humans
are destined to do the same.

I'm optimistic that, having been liberated from the myth of
intrinsic meaning, human beings will gain the ability to *make*
meaning instead, and that this unique ability will give us the
opportunity to disobey biology's commands.

The Future of Science, Religion, and Technology

ANTON ZEILINGER

Physicist, University of Vienna, and scientific director of the Institute of Quantum Optics and Quantum Information, Austrian Academy of Sciences.

I am optimistic about the future of science. Science as humanity's systematic endeavor of understanding nature is only a few centuries old. To believe that we have discovered the essentials of nature in such a short time is a sign either of arrogance or lack of imagination. So far, science has been guided by the (in my eyes) fallacious Cartesian divide between *res cogitans* and *res extensa*. It is wrong to believe that the world out there exists independent of our observation, but it is equally wrong to believe that it exists only because of our observation. We have to, and we will, find a completely new way of looking at the world which will transcend our present materialistic paradigm. After all, we have learned in quantum physics that all concepts of material existence evaporate. In the end, we are left with probability fields – probabilities of the results of observations. I am convinced that in science we have just started to scratch the surface. Our understanding of the world will be radically different from the understanding we have today.

I am optimistic about the future of religion. We will learn to shed the unessential dogmas, rules, definitions, and prejudices that religions have built up over centuries and millennia. We will learn that they have been created out of feelings of insecurity, out of an innate need of humankind to define and understand even the undefinable and non-understandable. I am convinced that in all major religions we will discover the essentials of what it means to be human. We will succeed in persuading church leaders and religious leaders to be more audacious, to open up to other views of the world and rely less on what they perceive to be their own access to truth. The present battle between science and religion will someday be seen as a battle between two unjustified positions. Science will never be able to prove that God does not exist, and religion will learn that its essence is far deeper than ephemeral questions like whether we were created by evolution or not. I believe that someday we will arrive at a coherent view of the world that will transcend both what today we call science and what today we call religion.

I am optimistic about the future of technology. Here too we have hardly scratched the surface. With quantum information technology, humankind for the first time is entering a field of technology that (according to what we know today) nature has not used in evolution. I am convinced that most of the technology of the future will be of that kind. New ideas will be created and new technologies will be invented that could come into existence only because we invented them – there is no other road to making them happen.

I believe in the future of humankind. As long as there are children, as long as there are people who look up at the night sky in sheer wonder, as long as there is music and poetry and the Mona Lisa – and old monasteries and young artists and fledgling scientists and all the other expressions of human creativity – I will remain optimistic.

Going Beyond Our Darwinian Roots

LEONARD SUSSKIND

Physicist, Stanford University; author of *The Cosmic Landscape*.

I am optimistic about the adaptability of the human brain to answer questions that evolution could not have designed it for. A brain that can rewire itself to visualize four dimensions, or the Heisenberg uncertainty principle, is clearly going way beyond what natural selection could have wired into it. This makes me optimistic that we may be able to go beyond our Darwinian roots in other ways.

A Secular Humanist Death

GEOFFREY MILLER

Evolutionary psychologist, University of New Mexico; author of *The Mating Mind*.

I'm optimistic about death. For the first time in the history of life on Earth, it is possible – not easy, but possible – for conscious animals like us to have a good death. A good death is a great triumph, something to be sought and accepted. Indeed, a good death should be recorded and broadcast as a moral example to us all.

What do I mean by a good death? I do not mean opiate-fueled euthanasia, or heroic self-sacrifice during flash-bang tactical ops, or a grudgingly tolerated end to a century of grasping longevity. I do not mean a painless, clean, or even dignified death. I mean a death that shows a gutsy, scientifically informed existential courage in the face of personal extinction. I mean a death that shows the world that we secular humanists really mean it.

There is, of course, no way to escape the hardwired fears and reactions that motivate humans to avoid death. Suffocate me, and I'll struggle. Shoot me, and I'll scream. The brain stem and amygdala will always do their job of struggling to preserve one's life at any cost.

The question is how one's cortex faces death. Does it collapse in mortal terror like a deflated soufflé? Or does it face the end of individual consciousness with iron-clad confidence in the persistence of virtually identical consciousnesses in other human bodies? My optimism is that in this millennium, well-informed individuals will have a realistic prospect of sustaining this second perspective right through the end of life, despite death's pain and panic.

When I die in fifty years, or next week, or whenever, here's what I hope I remember:

- My genes, proteins, neural networks, beliefs, and desires are practically identical to those sustaining the consciousness of 6 billion other humans and countless other animals, whose experiences will continue when mine do not.
- Since life must be common throughout the universe and resilient across time, such subjective experiences will continue not just on Earth in the short term but across many worlds, for billions of years.
- There is no spooky personal afterlife to fear or hope for, only

this wondrous diversity of subjectivity that trillions of individuals get to partake in.

- The more science one knows, the more certain and comforting this knowledge is.

These life lessons are, to me, the distilled wisdom of evolutionary psychology.

Many people resist this knowledge. They listen only to the hair-trigger anxieties of the amygdala – which constantly whispers 'Fear death, fear death.' They construct pathetic ideologies of self-comfort to plug their ears against such mortal terror. They nuzzle into reality's coarse pelt for a lost teat of supernatural succor. I call them the Gutless, because they aren't bright enough or brave enough to understand their true place in the universe. A whole new branch of psychology called terror-management theory studies the Gutless and their death-denying delusions.

A great ideological war is raging between the Godless (people like me, who trust life) and the Gutless – the talking heads of the extreme religious right, who fear death, fear the Godless, fear ongoing life in the future, when they no longer exist. I'm also optimistic about the outcome of this war, because people respect guts and integrity. People want moral role models who can show them how to live good lives and die good deaths. People want to believe that they are participating in something vastly greater and more wonderful than their solipsism. Science quenches that thirst far more effectively, in my experience, than any supernatural teat sought by the Gutless.

The War Between Science and Religion Will See New Light

MARCELO GLEISER

Professor of physics and astronomy, Dartmouth College; author of *The Prophet and the Astronomer: Apocalyptic Science and the End of the World.*

I'm optimistic that the debate or, should I say, war between science and religion will see new light. Right now, the fracturing seems to be worsening as further entrenchment occurs on both sides. Books (some of them from *Edge* colleagues) trashing religion as collective hallucination or idiotic superstition carry a simple message to people outside the sciences: We scientists are as radical as the religious extremists, as inflexible and intolerant as the movements we seek to exterminate by our oh-so-crystal-clear-and-irresistibly-compelling rationalizations.

Although I'm an atheist, I do not forget what is behind the power of religious thought: quite simply, hope. Life is tough, people suffer, and, rightly or wrongly, religion offers something for people to hold on to. Yes, it's crazy to believe in supernatural influences in the world and to devote your life to a God that seems to have vanished from the world for, by conservative estimate, at least 2,000 years. But scientists ought not to forget that most people need some sort of spiritual guidance, a kind of guidance that science (at least as it is taught today) cannot offer. Science has shown, and keeps showing, that we live in a cold, hard universe, completely indifferent to us and to life. And yet people love, die, connect, fight, and must come to some sort of inner peace or acceptance. What can science offer these people?

It is futile and naïve to simply dismiss the need people have for spirituality.

My hope is that people will begin to see science as a vehicle for mutual understanding and for respecting life. The more we study life and its mechanisms, the more we realize how rare it is, how precious it is. Sure, there may be life elsewhere, and it may even be intelligent. However, even if this is the case, odds are that we are still going to be stuck with ourselves, on this planet or in our solar neighborhood, for quite some time. Either science will teach us humility and respect for life or we will exterminate this most precious cosmic jewel. I am optimistic that scientists will teach people these lessons, instead of simply trying to rob them of their faith and offering nothing in return.

The First Coming

MARTIN E. P. SELIGMAN

Fox Leadership Professor of Psychology, University of Pennsylvania; author of *Authentic Happiness.*

I am optimistic that God may come at the end.

I've never been able to choke down the idea of a supernatural God who stands outside of time, a God who designed and created the universe. There is, however, an alternate notion of God relevant to the secular community – the skeptical, evidence-minded community that believes only in nature.

Isaac Asimov wrote a short story in the 1950s called 'The

Last Question.' The story opens in 2061, with Earth cooling down. Scientists ask the giant computer, 'Can entropy be reversed?' and the computer answers, 'Not enough data for a meaningful answer.' In the next scene, Earth's inhabitants have fled the white dwarf that used to be our sun for younger stars, and as the galaxy continues to cool they ask the miniaturized supercomputer, which contains all of human knowledge, 'Can entropy be reversed?' It answers, 'Not enough data.' This continues through more scenes, with the computer ever more powerful and the cosmos ever colder. The answer, however, remains the same. Trillions of years pass, and all life and warmth in the universe have vanished. All knowledge is compacted into a wisp of matter in the near-absolute zero of hyperspace. The wisp asks itself, 'Can entropy be reversed?'

'Let there be light,' it responds. *And there was light.*

Embedded in this story is a theory about God, based not on faith and revelation but on hope and evidence. Among the properties of God in the Judeo-Christian theory are omnipotence, omniscience, and benevolence, and he is seen as the creator of the universe. I think we need to give up the idea of a supernatural creator at the beginning of time. This is the most troublesome feature of the Judeo-Christian theory: It runs afoul of evil in the universe. If God is the designer, and also good, omniscient, and omnipotent, how come our part of the universe is so full of innocent children dying, of terrorism, of sadism? The creator idea also contradicts human free will. Why would an omnipotent and omniscient God have created a species endowed with free will so that this uppity species might defy Him? And who created the creator anyway?

There are crafty, involuted theological answers to each of

these conundrums. The problem of evil is allegedly solved by holding that God's plan is inscrutable: What looks evil to us isn't evil in God's inscrutable plan. The problem of reconciling human free will with the properties of God is a tough nut. Calvin and Luther gave up free will to save God's omnipotence. In contrast to this Reformation theory, modern 'process' theology holds that God started things off with a thrust toward increasing complexity (so far, so good). But mounting complexity entails free will and self-consciousness, and so human free will is a strong limitation on God's power. This theory of God gives up omnipotence and omniscience to allow human beings to possess free will. To circumvent 'Who created the creator?,' process theology gives up creation itself, by claiming that the process of becoming more complex is eternal – there was no beginning and there will be no end. So the process-theology God allows free will but at the expense of omnipotence, omniscience, and creation.

There is a different way out of these conundrums: It acknowledges that the creator idea is so antithetical to the other properties as to mandate jettisoning it. Importantly, this very notion – of God as the original creator – is what makes God so hard to swallow for the scientifically minded person. The creator is supernatural – an intelligent and designing being who exists before time and who is not subject to natural laws, a complex entity that occurs before the simple entities, thereby violating most scientific processes. Let the mystery of creation be consigned to the branch of physics called cosmology. Good riddance!

This leaves us with the idea of a God who had nothing whatever to do with creation but who is omnipotent, omniscient, and benevolent. Does this God exist?

Such a God cannot exist now, because we would be stuck once again with two of the same conundrums: How can there be evil if an existing God is omnipotent and benevolent, and how can humans have free will if an existing God is omnipotent and omniscient. So there was no such God and there is no such God now.

Consider the principle of 'nonzero,' which Robert Wright articulates in his 2000 book of the same name. Wright argues that the invisible hand of biological and cultural evolution ineluctably selects for the complex over the simple because positive-sum games have the survival and reproductive edge over zero-sum games, and that over epochal time more and more complex systems – fitfully but inevitably – arise. (Space does not allow me to expand on Wright's thesis and I must refer the justifiably unconvinced reader to his very substantial arguments.)

A process that selects for more complexity is ultimately aimed at nothing less than omniscience, omnipotence, and goodness. Omniscience is arguably the ultimate end product of science. Omnipotence is arguably the ultimate end product of technology. Goodness is arguably the ultimate end product of positive institutions. Altruism is selected for. So in the very longest run, we have a God who is not supernatural but who has acquired omnipotence, omniscience, and benevolence through natural processes. Perhaps – just perhaps – God comes at the end.

So I am optimistic that there may be, in the fullness of time, a First Coming. I am optimistic that this is the door through which meaning can enter our lives. A meaningful life is a life that joins with something larger than itself, and the larger that something is, the more meaning. I am optimistic that as individuals we can choose to be a tiny part of this process, which has as its ultimate end the bringing of a God who is omniscient, omnipotent, and benevolent.

A New Tool Leading Us Toward a Deep Understanding of Human Nature

FREEMAN J. DYSON

Physicist, Institute for Advanced Study; author of *Disturbing the Universe*.

I am generally optimistic, because our human heritage seems to have equipped us very well for dealing with challenges, from ice ages and cave bears to diseases and overpopulation. The whole species cooperated to eliminate smallpox. The women of Mexico reduced their average family size from seven to two-and-a-half in fifty years. Science has helped us to understand challenges and also to defeat them.

I am especially optimistic just now because of a seminal discovery that was made recently by comparing genomes of different species. David Haussler and his colleagues at UC Santa Cruz discovered a small patch of DNA they call HAR1, short for Human Accelerated Region 1. This patch appears to be strictly conserved in the genomes of mouse, rat, chicken, and chimpanzee, which means that it must have been performing an essential function that was unchanged for about 300 million years, from the last common ancestor of birds and mammals until today.

But the same patch appears grossly modified, with eighteen mutations, in the human genome, which means that it must have changed its function in the last 6 million years, from the common ancestor of chimps and humans to modern humans. Somehow that little patch of DNA expresses an essential difference between humans and other mammals.

We know two other significant facts about HAR1. First, it does not code for a protein but for RNA. Second, the RNA for which it codes is active in the cortex of the embryonic human brain during the second trimester of pregnancy. It is likely that the rapid evolution of HAR1 has something to do with the rapid evolution of the human brain during the last 6 million years.

The discovery of HAR1 marks the beginning of a new understanding of human evolution and human nature. I see it as a big step toward the fulfillment of the dream described in 1929 by Desmond Bernal, one of the pioneers of molecular biology, in his little book *The World, the Flesh and the Devil: An Enquiry into the Future of the Three Enemies of the Rational Soul.* Bernal saw science as our best tool for defeating the three enemies. 'The World' means floods and famines and climate changes. 'The Flesh' means diseases and senile infirmities. 'The Devil' means the dark irrational passions that plunge otherwise rational beings into strife and destruction. I am optimistic because I see HAR1 as a new tool leading us toward a deep understanding of human nature and toward the ultimate defeat of our last enemy.

Sometime in the 21st Century I Will Understand 20th-Century Physics

JERRY ADLER

Senior editor, *Newsweek*; coauthor (with Allan Gerson) of *The Price of Terror: Lessons of Lockerbie for a World on the Brink*.

I am optimistic that sometime in the 21st century I will understand 20th-century physics.

Not that I haven't tried before. Over the years (decades, really), I have enlisted some of the finest minds of the era to help me grasp relativity, quantum mechanics, and superstring theory: Richard Feynman, Sheldon Glashow, Paul Davies, Stephen Hawking – even Einstein himself, whose 1916 book *Relativity: The Special and General Theory* was the prototype for all subsequent efforts to explain the universe in words rather than equations. It marked the earliest appearance of that ubiquitous character, the man on the train – a faceless stick figure glimpsed through a window as he zoomed past at nearly the speed of light. Remarkably, we can observe him as he goes about his obsessive tasks: bouncing a ping-pong ball on a table or shining a flashlight at a mirror on the ceiling or holding up a clock for us to compare with the identical one we're holding. Many hours have I devoted to contemplating his inertial frame of reference and trying to reconcile it with my own as I stand motionless on the platform. I have engaged him in my own thought experiments, even conjuring a *Gedanken* companion who rides a train on the adjacent track. If they each pass my position at the same instant, traveling at three-quarters the speed of light in opposite directions, then their speed relative to each other is one-and-a-half times the speed of . . . Wait a second, that can't be right, can it?

What I'm up against here is a problem in translation. The laws of nature are written in equations, but I read only English. I have the same problem with anything written in French, of course, but I can accept a second-hand version of Proust more easily than one of Einstein or Heisenberg. My understanding of the world is not dependent on Proust the way it is on the double-slit experiment. Everything I know about the basic stuff of the universe – the very atoms I am made of, the gravity that glues me to my bed at night – I know second-hand, through the imperfect medium of language and metaphor. I don't even know what it means to 'solve' an equation in relativity or quantum mechanics. Here's an equation involving a man on a train that I *can* solve: If he leaves Chicago at 6:00 a.m. at seventy miles an hour, when will he pass someone who left St. Louis three hours earlier at fifty miles an hour? But when physicists 'solve' an equation, what emerges isn't a quantity, it's a new law of nature. How do they *do* that?

That's what keeps me awake at night, reading. It is much too late for me to go back and learn enough math to meet Einstein on his own terms, much less Heisenberg or Hawking. But I am sustained by optimism that someday I will transcend my own limitations, that I will achieve the conceptual breakthrough necessary to grasp relativity, quantum mechanics, and the rest of it on a deep level. Someday I will understand not just the epiphenomena of physics – the trains and the slits and the cats in boxes – but their mathematical essence. Their metaphysics. I'm optimistic. Really.

The Future of String Theory

GINO SEGRE

Professor of physics and astronomy, University of Pennsylvania; author of *Faust in Copenhagen: A Struggle for the Soul of Physics*.

I am optimistic about the future of our thinking regarding string theory and the early universe. Until fairly recently, I did not feel this way, since string theory seemed to be a community unto itself, albeit a very talented one. Controversy has created an important dialog, and strife has erupted. I think this is all to the good. The basis for the disagreement goes back thirty years.

A unified understanding, or so-called 'theory of everything,' has long been sought. The standard model of particle physics that emerged in the 1970s provided a very significant step forward but left undetermined some twenty parameters: the values of the six quark and six lepton masses, various couplings, and so on. Initially it was hoped that string theory, aside from a unification of forces with quantum gravity, would determine the values of these parameters. That dream has not been realized.

A significant group of theoretical physicists has now abandoned the dream. Pointing out that even string theory supports the view that an essentially infinite number of possibilities can be realized for a universe (the so-called 'landscape'), they maintain that we live in one of these choices – that is, the universe where the twenty or so parameters are fixed to be the values we observe. Other universes, with other values of the parameters, are continually emerging and dying; still others live by our side. However, we are limited in the possibility of observations and measurements to our own universe, so that, in a deep sense, the twenty parameters that determine our world are completely arbitrary. We would not exist if they were not what they are, but there is no further understanding of their values.

A second group maintains that abandoning the dream that set elementary-particle physics on its course a century ago – that of determining the forces and parameters of the subatomic world – is both premature and intellectually wrong. They maintain that this is not science.

There is an intermediate position that, understandably, has not been embraced vigorously by either side. Perhaps very few of the twenty or so parameters, some of the mass scales, correspond to the universe we live in, but the others are set by string theory or some future theory we have not yet discovered. This could be the case if, for example, the quark and lepton masses are calculable numbers that multiply a mass given by the particular universe we happen to live in. Thus both sides would be right. The numbers would be set by the theory and the mass scale by the choice of universe. I find the notion intriguing, but it may also be that both sides are wrong and that some other stunning synthesis will emerge.

So why am I optimistic? Because I believe that controversy, with clearly drawn opposing positions, galvanizes both sides to refine their opinions, creates excitement in the field for the participants, stimulates new ideas, and attracts new thinkers to the fray – and finally because it provides the public at large with an entrée into the world of science at the highest level, exhibiting heated arguments between great minds differing on questions vital to them. What could be more exciting?

Renewal of Science for the Public Good

LAWRENCE M. KRAUSS

Ambrose Swasey Professor of Physics and Astronomy, Case Western Reserve University; author of *Hiding in the Mirror: The Mysterious Allure of Extra Dimensions.*

I am optimistic that after almost thirty years of sensory deprivation in the field of particle physics, during which much hallucination (e.g., string theory) has afflicted theorists, we will finally, following the commissioning this year of the Large Hadron Collider in Geneva, obtain empirical data that will drive forward our understanding of the fundamental structure of nature, of its forces, and of space and time.

My biggest optimism is that the data will be completely unexpected, forcing revisions in all our carefully prepared ideas about what will supplant the standard model of elementary-particle physics. Since 1975 or so, every single experiment done at the microscopic forefront has been consistent with the predictions of the standard model, giving little or no direction to what lies behind it: What is the origin of mass? Why are there three families of elementary particles? Why are some quarks heavy? Why are neutrinos very light?

Yes, neutrino masses were discovered, but that was no big surprise, and no insight at all into their origin has been obtained thus far. With empirical data, theoretical particle physics may return to the days when the key to distinguishing good theory from bad was how many empirical puzzles the theory might resolve, rather than how fancy it looked.

I'm also optimistic that within what I hope will be my lifetime, we will unlock the secret of life and take our understanding of evolutionary biology back to that remarkable transition from nonbiological chemistry to biology. Not only will we be able to create life in the laboratory but we will be able to trace our own origins and gain insight into the question of how much life there is in the universe. We will surely discover microbial life elsewhere in our solar system – and I expect us to find that it's our cousin (from the same seed, if you will) rather than a true 'alien.'

Strangers in Our Midst

ROBERT SHAPIRO

Professor emeritus, senior research scientist, Department of Chemistry, New York University; author of *Planetary Dreams*.

I am optimistic about the prospect of detecting alternative life. All life that we know, as different as it may appear in size and shape, shares a common heritage at the biochemical level. From amoebas to zebras, familiar life is dominated by two types of large molecule – nucleic acids and proteins. This biochemical similarity, which extends to many other features as well, implies that we are all products of a single life-starting event.

If this event was extremely improbable, then Earth may be the only seat of life in an immense and barren universe. This picture would be little changed if our kind of life has drifted across empty space to fertilize our nearest planetary neighbors. As Jacques Monod commented: 'The universe was not pregnant with life nor the biosphere with man. Our number came up in the Monte Carlo game.'

As no firm evidence exists that supports or denies this package of gloom, we are not obliged to buy it. A different scientific position holds that the generation of life is written into the laws that govern the universe. If a limited set of environmental requirements are satisfied – a supply of useful energy, fertile material to absorb and use the energy, and a fluid medium to support the transaction – then life will emerge. In the words of biologist Stuart Kauffman: 'If all this is true, life is vastly more probable than we have supposed. Not only are we at home in the universe, but we are far more likely to share it with unknown companions.'

The issue can be settled by scientific exploration. The discovery of life-forms sufficiently different from our own to indicate a separate origin would tilt the debate decisively in favor of a fertile universe. The search for such life has traditionally been directed to extraterrestrial locales. Spacecraft have conducted preliminary surveys of Mars, Europa, Titan, and Enceladus and discovered that one or more of the necessary requirements have been met. As human beings have not yet traveled farther than the moon, the search for novel life-forms on these worlds is likely to be carried out by robots. If any creatures are encountered, then their biochemical characterization will also be conducted with programmed instruments, but weight limitations will constrain the versatility of the tools that can be landed on these distant worlds. The Viking missions of 1976 illustrate the ambiguities that can arise in such explorations. Even if encouraging data are returned to Earth, a sample-return mission would most likely be needed to settle a question of such magnitude. Safety considerations would make stringent quarantine measures mandatory for any returned samples.

Extensive planning and advance testing are needed to design a weight-limited apparatus capable of identifying alternative life. As astrobiology budgets are often under pressure, some delay is likely before launch of such an apparatus. Further, all of the listed sites except Mars require a number of years of travel time to bring the instrument package to its destination. Thus, even if the fertile-universe view is correct, many decades may pass before the issue is settled.

Fortunately, a new strategy has emerged that can provide much more rapid returns. One world exists that is known to have all of the capabilities needed to generate and sustain life. It is close at hand, so that any possible samples of alternative life could quickly be subjected to examination in depth, using the best instruments that science can provide. Human scientists would supervise the studies directly and modify them as needed. That world is Earth.

The suggestion that alternative life-forms might be found on our own planet runs directly into an entrenched paradigm. Biologists have characterized hosts of life-forms, particularly at the microbial level, and encountered the familiar nucleic-acid/protein-based system every time. Our type of life reigns on this planet. If alternative creatures ever existed, then surely they were eliminated in the intense combat of evolution. The fact that no such creatures have turned up despite the intense efforts expended in studying life on Earth reinforces this widely accepted conclusion.

Recently, however, two papers have challenged it, one by the physicist Paul Davies and the mathematician Charles Lineweaver and the other by the philosopher Carol Cleland

and the microbiologist Shelley Copley. Three of these authors and a number of other scientists interested in the question of extraterrestrial life (myself included) gathered at Arizona State University to discuss this possibility. A central conclusion that emerged was that alternative Earth life may simply have been overlooked because microbiological search techniques were targeted at our own kind of life. Many diverse cell-like objects can be observed when samples taken from soil or water are examined under the microscope. Only about 1 percent of them choose to multiply when conventional growth media are added to the mixture, and these colonies are the easiest to characterize. In some cases, newer techniques, based on nucleic-acid sequencing, have been used to identify additional species. Some of them represent early and unexpected branches from our presumed universal tree of life. The existence of truly different organisms in that mix – for example, organisms that lacked nucleic acids entirely and stored their hereditary information in some other way – was hardly considered. If such organisms exist, they would most likely be products of that speculative second origin.

How could such organisms have survived the competition of our robust nucleic-acid-based life? In one scenario, they may have preferred to dine on alternative foodstuffs not favored by familiar life, selecting, for example, arsenate in place of phosphate, unfamiliar amino acids, or mirror-image forms of conventional biomolecules. The most extreme example of this type may be the speculative mineral-based life-forms suggested by Scottish chemist Graham Cairns-Smith.

A different strategy would also allow alternative Earth life to flourish without direct competition. The organisms may have selected environments uninhabitable by conventional life.

Conventional terrestrial organisms have shown great versatility in adapting to extremes of acidity, temperature, dryness, saltiness, radiation, or other variables. Even so, their adaptability is not unlimited, and there are niches they cannot utilize. Yet organisms with a very different set of internal chemicals might find these to be ideal dwelling places. One such locality, mentioned at the conference, was Iron Mountain, California, from whose interior extremely acidic waters emerge.

Ironically (forgive the pun), a front-page story in the *New York Times* of December 23, 2006, derived from a paper published in *Science* a day earlier, reported the isolation of novel microorganisms from the waters of Iron Mountain. Their novelty arose from the record-breaking smallness of the cells rather than from a difference in their internal biochemistry sufficient to suggest a separate origin. Yet the group of California-based scientists that had made the discovery also noted the presence of 'rounded objects' that 'were not shown to contain DNA.' How could such objects – or others that may turn up if a deliberate search for biochemically novel organisms is made – be shown to be offspring of a second origin? My suggestion is that an inventory be made, as complete as possible, of their chemical contents. Many advanced instruments have been devised in recent years that can perform a microchemical analysis of tiny samples. No questions of instrument payload, robotic analysis, or sample return need be considered. If the analysis should reveal a chemical suite differing notably from those derived from conventional life and from the near random mixtures produced by abiotic processes, then we would have a strong indication that we have hit a scientific jackpot. Another approach to the identification of alternative life would involve the use of unorthodox culture media, toxic to conventional life, that induce the alternative organisms to grow. Strategies even more ingenious may emerge when the energies of additional scientists are turned to this question.

Of course, a thorough search of this planet may turn up nothing. My optimism is based on my particular outlook on the mechanisms involved in the origin of life – but that is another story.

Physics Will Not Achieve a Theory of Everything

FRANK WILCZEK

Physicist, MIT; Nobel laureate; author of *Fantastic Realities*.

I'm optimistic that physics will not achieve a Theory of Everything.

That might seem an odd thing to be optimistic about. Many of my physicist colleagues are inspired by the prospect of achieving a Theory of Everything. Some even claim they've already got it (acknowledging, to be sure, that perhaps a few i's remain to be dotted or a few t's crossed). My advice, dear colleagues: Be careful what you wish for. If you reflect for a moment on what the words actually mean, a Theory of Everything may not appear so attractive. It would imply that the world could no longer surprise us, had no more to teach us.

I don't buy it. I'm optimistic that the world will continue to surprise us in fascinating and fundamental ways. Simply writing down the laws or equations is a long way from being able to anticipate their consequences. Few physicists – and no sober ones – seriously expect future work in fundamental physics to exhaust, for example, neuroscience.

A less literal reading of 'Theory of Everything' is closer to what physicists who use it mean by it. It's supposed to be a theory not of everything but of 'everything fundamental.' And 'fundamental' is also being used in an unusual, technical sense. A more precise word here might be 'irreducible.' That is, the physicists' Theory of Everything is supposed to provide all the laws that can't be derived logically, even in principle, from other laws. The structure of DNA surely emerges – in principle – from the equations of the standard model of particle physics, and I strongly suspect that the possibility of mind does too. So those phenomena, while they are vastly important and clearly fundamental in the usual sense, aren't fundamental in the technical sense, and elucidating them is not part of a Theory of Everything.

We're about to enter a new golden age in fundamental physics. The Large Hadron Collider (LHC), which should begin to operate at CERN, near Geneva, in the summer of 2007, will probe the behavior of matter at energies higher than ever accessed before. There is no consensus about what we'll find there. I'm still fond of a calculation that Savas Dimopoulos, Stuart Raby, and I did in 1981. We found – speaking roughly – that we could unify the description of fundamental interactions (gauge unification) only within an expanded version of relativity, which includes transformations of spin (supersymmetry). To make that dual unification, we had to bring in new particles – which were too heavy to be observed at the time but ought to be coming into range at the LHC. If they do exist, we'll have a new world of phenomena to discover and explore.

The astronomical riddle of dark matter could well be found there. Several competing ideas are in play as well. The point is that whatever happens, experimenters will be making fundamental discoveries that take us by surprise. That would be impossible if we had a Theory of Everything in the sense just described – that is, of everything fundamental.

In recent months, a different, much weaker notion of what a Theory of Everything might accomplish has gained ground, largely inspired by developments in string theory. In this concept, the TOE provides a unique set of equations, but those equations have many solutions, which are realized in different parts of the universe. One speaks, instead, of a multiverse, composed of many domains, each forming a universe in itself, each with its own distinctive laws. Thus, even the fundamental – i.e., irreducible – laws are beyond the power of the TOE to supply, since they vary from universe to universe. At this point, the contrast between the grandeur of the phrase 'Theory of Everything' and the meager information delivered becomes grotesque.

The glamour of the quest for a Theory of Everything, or a Final Theory, harks back to Einstein's long quest for his version, a Unified Field Theory. Lest we forget, that quest was fruitless. During his great creative period, Einstein produced marvelous theories of particular things: Brownian motion, the photoelectric effect, the electrodynamics of moving bodies, the equality of inertial and gravitational mass. I take inspiration from the early Einstein, the creative opportunist who consulted nature, rather than the later 'all-or-nothing' romantic who tried (and failed) to dictate to it. I'm optimistic that it will continue to surprise me and my successors for a long time.

Bullish on Cosmology

PAUL STEINHARDT

Physicist; Albert Einstein Professor of Science, Princeton University; coauthor (with Neil Turok) of *Endless Universe: A New History of the Cosmos.*

I am optimistic that there will be a historic breakthrough in our understanding of the universe in the next five years – one that will be remembered as among the most significant of the millennium. I would also give better-than-even odds that there will be more than one discovery of this magnitude.

My optimism is sparked by a remarkable coincidence: the simultaneous maturing of several unrelated technologies, each of which could open a new window on the cosmos. Historically, every new technology is a harbinger of great discovery; thus, at least a handful of major advances will occur within just five years:

Direct detection of dark matter

After decades of gradual progress, physicists will finally build the first detectors sensitive enough to detect dark-matter particles directly, if they consist of WIMPs (weakly interacting massive particles, as many physicists suspect.

Discovery of the nature of dark energy

Although their names sound similar, the only quality that dark matter and dark energy have in common is that they are both invisible. Dark matter consists of massive particles that gravitationally attract one another and clump into clouds that seed the formation of galaxies. Dark energy is gravitationally self-repulsive, so it tends to smooth itself out. When it is the dominant form of energy, as it is today, dark energy causes the expansion of the universe to speed up. The composition of dark energy is one of the great mysteries of science, with profound implications for both fundamental physics and cosmology.

Over the next five years, arrays of novel wide-field telescopes will be constructed that are programmed to rapidly scan large fractions of the sky in search of phenomena that vary rapidly with time. The arrays will be used to search for distant supernovas (exploding stars), whose brightness and colors can be used to judge the distance and recessional speed of their host galaxies. From these measurements, astronomers can measure precisely the accelerated expansion of the universe, a primary means of distinguishing different theories of dark energy.

At the same time, in the laboratory, physicists will be trying to detect changes in the gravitational force when masses are placed at close proximity or tiny changes in the strength of the electromagnetic force with time, other effects predicted by some theories of dark energy. These measurements will significantly narrow the candidates for dark energy, perhaps identifying a unique possibility.

Exploring the Big Bang and the origin of the large-scale structure of the universe

The conventional wisdom is that the universe sprang into existence 14 billion years ago in a Big Bang and that a period of exponentially rapid inflationary expansion accounts for its large-scale structure. However, the last decade has seen the emergence of alternative possibilities, such as the cyclic model of the universe.

In the cyclic model, the Big Bang is *not* the beginning but, rather, an event that has been repeating every trillion years, extending far into the past. Borrowing ideas from string theory, the cyclic model proposes that each bang is a collision between our three-dimensional world and another three-dimensional world along an extra spatial dimension. Each bang creates new hot matter and radiation that begins a new period of expansion, cooling, galaxy formation, and life, but space and time exist before and after the bang.

The large-scale structure of the universe and the pattern of galaxies are set by events that occurred about a cycle ago, before the bang, just as events occurring today are setting the structure for the cycle to come. Although the inflationary and cyclic pictures predict distributions of galaxies, matter, and radiation that are indistinguishable, their predictions for the production of gravitational waves in the early universe are exponentially different.

Gravitational waves are ripples in space produced during inflation, or near the beginning of a new cycle, that propagate through the universe and distort space, like undulations traveling through Jell-O. These cosmic gravitational waves are too weak to be detected directly, but experimental cosmologists throughout the world are mounting ground- and balloon-based experiments to search for their imprint on the polarization pattern of cosmic microwave background radiation produced in the first 380,000 years after the Big Bang. The results will affect not only our view of our cosmic origin but our future as well. The conventional Big Bang inflationary theory predicts that our universe is headed toward the cold oblivion of eternal expansion – a whimper – but the cyclic model predicts a new, hot Big Bang.

Direct detection of gravitational waves

The first window on the universe using something other than electromagnetic waves could be open within the next five years. After decades of developments, the LIGO (Laser Interferometer Gravitational-wave Observatory), with one detector in Livingston, Louisiana, and one in Hanford, Washington, has a plausible chance of directly detecting gravitational waves, beginning a new era in astronomy. The observatory is designed to detect stronger gravitational waves than those produced in the early universe – such as waves generated by the violent collision of neutron stars and black holes in our own galaxy. However, this frontier is so fresh and unexplored that there could well be unanticipated cosmic sources of strong gravitational waves to be discovered that could cause us to reassess our understanding of the universe.

Breakthroughs in fundamental physics and direct production of dark matter

The Large Hadron Collider at the European Organization for Nuclear Research (CERN) in Geneva, Switzerland, is set to begin operation this year. This facility consists of a powerful particle accelerator that will reproduce collisions of the type that occurred within the first pico-second after the Big Bang, carrying the investigation of fundamental physics over an important energy threshold, where new phenomena are anticipated. For example, physicists hope to discover a spectrum of new 'super-symmetric' particles, confirming a key prediction of string theory, and also WIMPs that may comprise the dark matter.

The impact will be profound. Currently, we understand the composition of less than 5 percent of the universe; we do not understand how space, time, matter, and energy were created; and we cannot predict where the universe is headed. In the next five years, we may witness the historic resolution of one or more of these issues. I have my personal bet on what the individual outcomes will be, but the only prediction I will reveal here is that with the opening of so many new windows on the cosmos, we are sure to discover something unanticipated and astonishing.

The Return of the Discipline of Experiment Will Transform Our Knowledge of Fundamental Physics

LEE SMOLIN

Founding member and research physicist, Perimeter Institute for Theoretical Physics, Waterloo, Ontario, Canada; author of *The Trouble with Physics*.

In science, as in politics, it seems that Niles Eldredge and Stephen Jay Gould's metaphor of punctuated equilibrium holds. When progress happens, it happens fast and the whole culture vibrates with the excitement of it. We have had a bit too much equilibrium lately, of disappointed expectations following as a natural consequence of unwisely reduced ambitions. But I am optimistic that the next decades will see breakthroughs in key problems on which we now seem stuck. In physics, new experiments including the LHC, AUGER, GLAST, PLANCK, LIGO, and others are likely to transform our knowledge of fundamental physics and end the long period when theory sought to progress without the discipline of experiment. Very likely we will be surprised and humbled by what is seen, but this will be followed by rapid progress as new ideas are quickly invented to explain the surprising data.

How can I be optimistic without knowing what direction science will take? This is exactly the point. There are two kinds of optimism, the optimism of people who think they know the future and the optimism of people who believe the future will be more interesting and, if always imperfect, more wonderful than they can imagine. I am of the second kind. The first kind sometimes comes along with a belief that time and change are illusions and that the world is evolving toward an eternal time-less state of perfection. This is the optimism of religious fundamentalists and orthodox Marxists, and one sees it reflected also in the cosmologies in which our evolving universe is just a transient fluctuation in an otherwise permanent state of thermal equilibrium. The opposite kind of optimism lies behind the evolutionary theorists who believe that the world is so intricate that the simplest mechanism that could predict the future of life and the cosmos is the universe itself. If we are the first kind of optimist, we seek to transcend the complexi-ties of life to discover something eternal behind it, something like the imagined view of God. If we are the second, we seek to live and think within the swirl of life; we aim for comprehension and wisdom but have no illusions of transcendence or control.

People Will Increasingly Value Truth (over Truthiness)

LISA RANDALL

Professor of physics, Harvard University; author of *Warped Passages: Unraveling the Mysteries of the Universe's Hidden Dimensions.*

Optimism is an 'ism' like any other. People reading these pages should recognize the responses as the hopeful beliefs they are. With this caveat, I'm optimistic that people will increasingly value truth (over truthiness). After its recent digressions into beliefs and images dominating current thought, I'm anticipating that society will increasingly recognize and understand the value of knowledge. People will want to make their own critical judgments, know more facts, and stop turning to questionable authorities or visual media for their education. I don't necessarily think everyone will do so, but I'm optimistic that the ones who do won't remain a silent minority.

Part of my optimism stems from my experiences talking and writing about the future of particle physics to the public and the surge of interest I've found when people realize how much they can learn and understand. All of us in the particle physics community are eagerly awaiting the Large Hadron Collider, a proton-proton collider being built in Switzerland that will turn on in 2007. I'm optimistic (with calculations supporting my optimism) that this machine will tell us about the nature of mass and explain to us the weakness of gravity relative to the other known elementary-particle forces. I'm optimistic that we'll learn something truly new and exciting about the fundamental nature of matter and our world – maybe something as exciting as extra dimensions of space, or perhaps something no one has even thought about yet. Whatever the results will be, the LHC gives us reason to be optimistic.

Physics Will Flourish Once More

CHARLES SEIFE

Professor of journalism, New York University; former journalist for *Science* magazine; author of *Zero: The Biography of a Dangerous Idea.*

It's easy to be a pessimist. These are dark times for physics in the United States. One by one, national laboratories are shuttering their high-energy physics experiments; within a few years there won't be a single U.S. accelerator exploring the energy frontier. As NASA squanders billions and billions of dollars on the International Space Station and on lunar exploration, it is tearing the guts out of its other programs – the ones that provide actual scientific discoveries.

Physics is a transplant to the United States. Before 1900, you could count the number of great American physicists on one hand. A few decades later, the U.S. had become the premier power in theoretical and experimental physics, thanks to refugees from Hungary, Austria, Germany, Italy, Denmark, and all across Europe. The transplant took root and flourished.

Even though the future is dimming for American physics, there is room for optimism; the prospects for major discoveries are the brightest they've been in years. We are in the midst of a cosmological revolution; we are beginning to understand the physical laws that governed the early universe. Before the end of the decade, European experiments, such as the Planck satellite and the Large Hadron Collider at CERN, will allow physicists to delve deeper than ever before into the story of the infant cosmos. While physics in the United States is withering, there will be fertile soil where a transplant can take root once again.

The Optimism of Scientists

KARL SABBAGH

Writer and television producer; author of *The Riemann Hypothesis*.

To ask, 'What am I optimistic about?' is rather like asking, 'What am I tall about?' or 'What am I English about?' For me, optimism is a personal characteristic, rather than an attitude to be applied to some things and not others. Fortunately it is a characteristic that many scientists have and others acquire, and I am optimistic that this optimism will continue to be a signal human characteristic. Without optimism, why would anyone embark on the complex and interrelated series of steps that makes up any scientific experiment, let alone huge enterprises like the Manhattan Project or the Apollo Program? And faced with disasters like *Challenger* and *Columbia* and the results of inquiries into how they happened, how could anyone have the faith to continue, without optimism?

The Large Hadron Collider at CERN is perhaps the greatest testament to this optimism. Conceived decades ago, absorbing $2.5 billion in a collaboration among more than forty countries, and designed to accelerate invisible particles to 99.999999 the speed of light in order to create a theoretical entity, the Higgs boson, for which no evidence exists – if this is not a triumph of optimism over realism, I don't know what is.

And I believe this optimism is more than just logical and reasoned inference from previous researches. Scientists are optimistic about science in general as a tool for discovery. They believe that the methods of science will produce valid results. They believe that whatever aspect of the universe – even one never previously explored – they turn their attention toward, they can design experiments and carry out observations that will be valid and provide sustainable increments in our understanding. Is this optimism unique to science? I believe it is. Few people have a comparably strong faith in the future benefits of politics or economics or art or philosophy or technology. Some favor capitalism, others socialism; some favor nuclear power, others renewable energy; some believe in a wide-ranging humanistic education, others believe in vocational training; some believe in nationalism, others in internationalism. But scientists believe in science – that is an indication of their optimism.

What Lies Beyond Our Cosmic Horizon?

ALEXANDER VILENKIN

Director of the Institute of Cosmology, Tufts University; author of *Many Worlds in One*.

There is a limit to how far we can see into the universe. Our cosmic horizon is set by the distance traveled by light since the Big Bang. Objects more distant than that cannot be observed, because their light has not yet reached Earth. But the universe does not end at the horizon, and the question is, What lies beyond? Is it more of the same – more stars, more galaxies? – or could it be that remote parts of the universe are very different from what we see around here? I am optimistic that we will be able to answer this question and understand the structure of the universe as a whole, even though we can observe only a small part of it.

Until recently, cosmologists made the simplest assumption – that the universe is homogeneous – that is, that it looks every-where more or less the same. But recent developments in cosmology and particle physics have led to a drastic revision of this view and to a heated debate about the future of our science. According to the new worldview, most of the universe is in the state of explosive, accelerated expansion called inflation. In our local region (that is, the observable universe), inflation ended 14 billion years ago, and the energy that drove the expansion went to ignite a hot fireball of elementary particles. This is what we call the Big Bang. Other Big Bangs constantly go off in remote parts of the universe, producing regions with diverse properties. Some of these regions are similar to ours, while others are very different.

The properties of any given region are determined by the quantities we call constants of nature. These include particle masses, Newton's constant (which controls the strength of gravity), and so on. We do not know why the constants in our region have their observed values. Some physicists believe that these values are unique and will eventually be derived from some fundamental theory. However, string theory, which is at present our best candidate for the fundamental theory of nature, suggests that the constants can take a wide range of possible values. Regions of all possible types are then produced, in the course of eternal inflation. This picture of the universe – or multiverse, as it is called – explains the long-standing mystery of why the constants of nature appear to be fine-tuned for the emergence of life. The reason is that intelligent observers exist only in those rare regions in which, by pure chance, the constants happen to be just right for life to evolve. The rest of the multiverse remains barren, but no one is there to complain about that.

Many of my physicist colleagues find this multiverse picture alarming. Since all those regions with different values for the constants are beyond our horizon, how can we verify that such regions really exist? Is this science – to talk about things that can never be observed? In my view, it *is* science, and there are good reasons to be optimistic about the new picture. If the constants vary from one part of the multiverse to another, their local values cannot be predicted with certainty, but we can still make *statistical* predictions. We can try to predict what values of the constants are most likely to be observed. One such prediction, that the vacuum should have a small nonzero energy, has already been confirmed. We have only started along this path, and formidable challenges lie ahead. I believe, however, that what we are facing now is not a dead end, as some people fear, but the beginning of a new era – the exploration of the multiverse.

We're Not Insignificant After All

MAX TEGMARK

Associate professor of physics, MIT; researcher, Precision Cosmology.

When we're gazing up on a clear night, it's easy to feel insignificant. Since our earliest ancestors admired the stars, our human egos have suffered a series of blows. For starters, we're smaller than we thought. Eratosthenes showed that Earth was larger than millions of humans had thought, and his Hellenic compatriots realized that the solar system was thousands of times larger still. Yet for all its grandeur, our sun turned out to be merely one rather ordinary star among hundreds of billions in a galaxy that, in turn, is merely one of billions in our observable universe, the spherical region from which light has had time to reach us during the 14 billion years since our Big Bang. There are probably more (perhaps infinitely many) such regions. Our lives are small temporally as well as spatially: If this 14-billion-year cosmic history were scaled to one day, then 100,000 years of human history would be 4 minutes and a 100-year life span would be 0.2 seconds. Further deflating our hubris, we've learned that we're not that special either. Darwin taught us that we're animals, Freud taught us that we're irrational, machines now outpower us, and just last December Deep Fritz outsmarted our chess champion, Vladimir Kramnik. Adding insult to injury, cosmologists have found that we're not even made out of the universe's most prevalent substance.

The more I learned about all this, the less significant I felt. Yet in recent years I've turned more optimistic about our cosmic significance. I've come to believe that advanced evolved life is very rare yet has huge growth potential, making our place in space and time remarkably significant.

The nature of life and consciousness is a hotly debated subject. My guess is that these phenomena can exist much more generally than in the carbon-based examples we know of. I believe that consciousness is, essentially, the way information feels when being processed. Since matter can be arranged to process information in numerous ways of vastly varying complexity, this implies a rich variety of levels and types of consciousness. The particular type of consciousness that we subjectively know is thus a phenomenon that arises in certain highly complex physical systems that input, process, store, and output information. Clearly, if atoms can be assembled to make humans, the laws of physics also permit the construction of vastly more advanced forms of sentient life. Yet it's likely that such advanced beings can come about only in a two-step process: First, intelligent beings evolve through natural selection; then they choose to pass on the torch of life by building more advanced consciousness that can further improve itself.

Unshackled by the limitations of our human bodies, such advanced life could rise up and eventually inhabit much of our observable universe. Science fiction writers, AI aficionados, and transhumanist thinkers have long explored this idea, and to me the question isn't if it *can* happen but if it *will* happen.

My guess is that evolved life as advanced as ours is very rare. Our universe contains countless other solar systems, many of which are billions of years older than ours. Enrico Fermi pointed out that if advanced civilizations have evolved in many of them, then some have a huge head start on us – so where are they? I don't buy the explanation that they're all choosing to keep a low profile. Natural selection operates on all scales, and as soon as one life-form adopts expansionism (sending off self-replicating interstellar nanoprobes, say), others can't afford to ignore it. My personal guess is that we're the only life-form in our entire observable universe that has advanced to the point of building telescopes. Let's explore that hypothesis. It was the cosmic vastness that made me feel insignificant to start with, yet those galaxies are visible and beautiful to us – and only us. It is only we who give them any meaning, making our small planet the most significant place in our observable universe.

Moreover, this brief century of ours is arguably the most significant one in the history of our universe: The one when its meaningful future gets decided. We'll have the technology either to self-destruct or seed our cosmos with life. The situation is so unstable that I doubt we can dwell at this fork in the road for more than another hundred years. If we end up going the life route rather than the death route, then in a distant future our cosmos will be teeming with life, all of which can be traced back to what we do here and now. I don't know how we'll be thought of, but I'm sure that we won't be remembered as insignificant.

Coraggio, Domani Sarà Peggio!

GEORGE F. SMOOT

Cosmologist, Lawrence Berkeley National Laboratory, professor of physics, University of California at Berkeley; Nobel laureate; coauthor (with Keay Davidson) of *Wrinkles in Time*.

'Courage, for tomorrow will be worse!' For some reason, this has been a mantra of mine since I first heard it, many years ago, issuing from a tenor in some Italian opera or other. The implications are very large.

A careful assessment and years of experience show that the long-term future is bleak: Entropy will continue to increase, and a heat death (actually a misnomer, as it means the degradation of usable energy in a dull, cooling, worthless background of chaos) is the inevitable fate of the world. This is the fate that awaits us if we manage to work our way past the energy crisis that looms as the sun runs out of fuel and, in its death throes, expands as a red giant star, likely to engulf Earth after boiling away the seas, before it collapses to a slowly cooling cinder, eventually to leave the solar system in cold darkness.

This energy crisis will eventually spread to the whole Milky Way Galaxy, which will use up its available energy resources on a time scale of roughly ten times the present 14-billion-year lifetime of our observed universe. In that same time, the accelerating expansion of the universe will continually reduce what we can observe and potentially access, and in the distant future only the cinders of stars will remain. Argument goes on as to whether a sufficiently advanced intelligent society could manage to live (continue to have experiences, process new information, create new things) indefinitely in such an environment, taking on the most carefully constructed and extreme measures that are physically possible. The chances of success look relatively low for even the most optimally managed and intelligent society.

Given this, and the history of human society cooperatively plundering the resources of a beautiful planet with abundant yet finite resources, who can possibly be optimistic about the long-term future of humanity? How many examples do we have of humans proactively addressing global problems in an efficient way and with enlightened selflessness? History tells us that humans have generally been engaged in warfare, exploitation for personal gain, and religious strife. Issues are generally not addressed until they become serious crises – and often not even then. This is exemplified by various episodes of genocide, large-scale pollution, and ecological devastation – phenomena that are often related.

Against this background, the rise of culture and science is remarkable – that is, until we recognize their usefulness. In our modern world, it is clear that material rewards and political power accrue to those societies that have the most scientific and technological knowledge and an educated workforce with the ethos of hard, productive work. Such societies enjoy economic and military success and large tax revenues.

Is it this culture and knowledge that offers us the hope of being at least as successful as the dinosaurs, which dominated Earth for nearly a hundred times as long as humans have? Could humans have shared the planet with reptiles? How about intelligent machines or advanced evolved beings? Could they share the planet with us?

Leaving cultural and religious value systems aside, let us move on to a realistic assessment of this essay's title. *Will* tomorrow be worse? Every physical process in the universe follows the second law of thermodynamics: That is, in every process, entropy (a measure of disorder that equals loss of information and usefulness) will tend to increase for the universe as a whole. No process decreases the entropy of the universe; only completely reversible processes leave it unchanged. All living things, and all man-made machines, operate by means of processes that increase the entropy of the universe. One cannot live by the dictum 'Do no harm.' The best one can aspire to is 'Do minimal damage.'

I was often bothered by this inevitable conclusion and tried to determine whether one could write a great work of literature, create art, or make (optimally) a great scientific discovery and thus leave the world better than one found it. Each time I worked out an example, the effect was negligible, no matter how valuable it was, compared to the damage done by mere existence. The only discoveries that would make a difference either called for repealing or avoiding the laws of probability or for making a whole new universe. Perhaps the discovery of extra dimensions will allow some leeway in what otherwise seems an inescapable doom after a long period of degradation of the universe. We face a continuous downward spiral of no return. This is not a moral or ethical statement, only an engineering evaluation – though it is some indication of original sin. Even living one's life as a vegetarian, eating only fruit falling into one's hand from a willing plant, and being kind to other living entities is not enough, for those other entities are worsening the universe for the sake of making a little more order for themselves.

With such sure knowledge of one's impending demise and that of all humanity, how can one face each day with any hope? In science, it is a central tenet to be skeptical, to question – so as certain as all this seems, one can cling to a shred of hope that some trapdoor will be found and opened, thanks to the infinite flexibility of humankind and our incredible ingenuity under pressure. But I suspect that the odds of winning the lottery are higher. Much higher.

In the face of this bleak scientific forecast and the inevitable erasure of any achievement, how is one to find a more optimistic philosophy? How about appealing to the many positive aspects of the human spirit, and what can result from actualized potential. Courage, optimism, hope, and the firm belief that humans have the potential for greatness can help us to navigate through what feels like a desperate spiral toward nothingness. In spite of vast cosmic mayhem, people still do great and glorious deeds: They build civilizations, cure disease, fight for social justice. These are good and inspiring beliefs and can provide hope.

I take a more Spartan approach: One simply depends upon courage to go forward to accomplish things, to build a society and a smoothly functioning culture. Going forth is fulfilling, satisfying, important to living; it provides a connection to past and future human generations, perhaps reaching back even to our one-celled-organism ancestors and ahead to whatever may arise in the distant future.

I personally am an optimist in many ways, and I will remain so. I invest time and effort in my work and my health. I fund my retirement plan generously with the idea that I will live a long, healthy, and productive life. I write articles and spend much of my days training and educating the next generation. Why do I do this? Because I believe that investing in yourself and in others is the hope for the future. The genius works produced by the great writers, activists, artists, musicians, and scientists are proof of the power of the human spirit and what can result when innate talent is combined with hard work and devotion. I look forward to the immediate future, as a part of the long human slog toward a better culture and a better society, in spite of the constant flux of misguided craziness.

Progress Is Surprisingly Durable

JAMES O'DONNELL

Classicist, cultural historian, provost of Georgetown University; author of *Augustine: A New Biography.*

Anna Karenina famously begins with the line, 'Happy families are all alike; every unhappy family is unhappy in its own way.' A little less famously and a great deal more astutely, Vladimir Nabokov turned the line on its head at the opening of *Ada*: 'All happy families are more or less dissimilar; all unhappy ones are more or less alike.' I'm with Nabokov, and that's why I can be an optimist.

Of course, in the long run optimism is impossible. Entropy is unforgiving: Even a historian knows that. And history repeats itself. The same stupidities, the same vengeances, the same brutalities are mindlessly reinvented over and over again. The study of history can help the educated and the wise avoid the mistakes of the past but, alas, it does nothing for the numbskulls.

Yet the study of the past and its follies and failures reveals one surprising ground for optimism. In the long run, the idiots are overthrown, or at least they all eventually die, whereas creativity and achievement are unique, exciting, liberating – and abiding. The discoveries of scientists, the inventions of engineers, the advances in the civility of human behavior are surprisingly durable. They may be thwarted or attacked. At any given moment, it may seem that the cause of, say, women's rights is beleaguered in too many places in the world. But the *idea* of women's equality with men is not going to vanish. Too few students may master the natural sciences, but the understanding enshrined in Newton's laws of motion and the calculus are not going to vanish. Too many people may eat and smoke their way to early graves, but the accurate understanding of the mechanisms of the human body and how they can be healed and repaired and kept healthy – that's not vanishing, either.

After all, we started out in the African savannah, trying to run fast enough to catch up with things we could eat and stay away from things that could eat us. Our natural destiny is to squat in caves and shiver, then die young. We decided we didn't like that, and we figured out how to do better. Even if the numbskulls get their way and we wind up back in a cave, we will remember – and we won't be in the cave for long. We do not remember everything, and there are losses. But we turn out to be a stubbornly smart, resilient, and persistent species, and we do not forget the most important things.

The Situational Focus

PHILIP G. ZIMBARDO

Professor of psychology emeritus, Stanford University; author of *The Lucifer Effect: Understanding How Good People Turn Evil.*

In trying to understand human behavior that violates our expectations, there is a learned tendency to 'rush to the dispositional.' We seek to explain behavior in terms of the personal qualities of the actor. In individualistic cultures, this means searching for genetic, personality, or pathological characteristics that can be reasonably linked as explanatory constructs. It also has come to mean discounting or ignoring aspects of the behavioral context – situational variables – that may be significant contributors to the behavior. Dramatists, philosophers, and historians, as well as clergy and physicians, all tend toward the dispositional and away from the situational in their views of human nature.

Social psychologists have been struggling to modify this bias toward inner determinants of behavior by creating a large body of research highlighting the importance of outer determinants. Rules, responsibility, anonymity, role-playing, group dynamics, authority pressures, and more have been shown to have a dramatic effect on individual behavior across a variety of settings.

The social psychologist Stanley Milgram's classic demonstration of blind obedience to authority showed that most ordinary Americans would follow orders given by an authority even if it led to severely harming an innocent person. My Stanford prison experiment extended this notion of situational power to demonstrate that institutional settings – prisons, schools, businesses – exert strong influences over human behavior. Nevertheless, the general public (and even intellectuals from many fields) still buys the dispositional and dismisses the situational as mere mitigating circumstance.

I am optimistic that this bias will be rebalanced in the coming year, as new research reveals that the situational focus is to an enhanced public-health model as the dispositional is to the old medical model in trying to understand and change behavior of people in communities. The focus of public health on identifying vectors of disease can be extended to systemic vectors of health and success in place of individual ambition and personal failure or success.

This analysis will be important in meeting the challenges posed by international terrorism through new efforts to build community resilience instead of focusing on individual coping. It will also change the blame game of those in charge of various institutions and systems – from identifying the 'few bad apples' to actively trying to understand how the apple barrel is corrupting good apples. I have shown how this dispositional thinking operated in analyzing the causes of the abuse and torture at Abu Ghraib by the military and civilian chains of command. Dispositional thinking is no different than the search for evil by identifying and destroying the 'witches' in Salem. Although the foundations of such thinking run deep and wide in most of us, I am optimistic that we will acquire a more balanced perspective on how good people may turn evil and bad people can be guided toward good.

The Women of the 110th Congress

ROGER BINGHAM

Cofounder and director of the Science Network; neuroscience researcher, Center for Brain and Cognition, UC San Diego; coauthor (with Peggy La Cerra) of *The Origin of Minds: Evolution, Uniqueness, and the New Science of the Self.*

I am optimistic about the record number of women who came to Washington last January as members of the 110th Congress – sixteen senators and seventy-one representatives. Only twenty years ago, there were just two female senators and twenty-three representatives. Now the Speaker of the House is a woman, second in the line of succession to the presidency.

Why do I think this ascent of women is cause for optimism? Because I believe we need a lot more social smarts – particularly empathy – in the corridors of power and the brains of our political leaders, and current evidence indicates that in general the female brain is intrinsically a more proficient empathizing device.

One example: Not long ago, Tania Singer (then at University College London, now at the University of Zurich) published a study on the neural processes underlying empathy. As one rather lurid headline described the work: 'Revenge Replaces Empathy in Male Brain. Watching bad guys suffer lights up the mind's reward centers for men.' The experiment involved imaging the brains of a group of male and female volunteers while they played a monetary investment game based on trust. A few of the players were actors, planted to play fairly or unfairly. Subsequently, the actors were zapped with a mild electric shock while the other players watched. When an actor who had played fair was shocked, both female and male volunteers felt their pain: fMRI images showed empathic activation of pain-related areas of their brains, specifically in the insular and anterior cingulate cortices. But when the unfair actor was shocked, the sexes reacted differently. The females still showed a surge of empathy; the males didn't. Instead, their reward areas lit up. The German word for this feeling – *Schadenfreude* – roughly translates as taking pleasure in the misery of others. The researchers' conclusion, based on the scanning and the comments on post-experiment questionnaires, was that 'Men expressed a stronger desire for revenge than women.'

Twenty years ago, in a PBS program called *The Sexual Brain*, I explored male-female differences. It was quite clear then (and has since become even more evident) that anatomically, chemically, and functionally, the brains of men and women have significant differences. To what extent these variations drive, or map onto, differences in cognition and behavior remains controversial (ask Larry Summers). But we know, for example, that men have a proportionally larger amygdala than women and women have a proportionally larger prefrontal cortex.

There's evidence, too, that women have a proportionally larger anterior cingulate cortex (ACC) and insula than men. From recent brain-imaging studies, we know that these areas are important components of the circuitry underpinning our ability to process complex social emotions, 'read' the faces and minds of the people we encounter as we navigate through social space, understand and predict their behavior, make moral judgments and empathize. (Of course, everything I am saying should have the usual scientific safety net of caveats: These differences exist at the level of populations, not individuals; cultural factors play an important role; and so on.)

So what does this have to do with my optimism about the women of the 110th Congress? Would women govern differently than men?

That's exactly the question that Geneva Overholser posed in the *New York Times* in 1987, just after Margaret Thatcher won a new term as British prime minister. Overholser noted that Thatcher – like the few other women who were national leaders – operated in a male-dominated political arena and basically behaved no differently in office from men. 'Would that be true,' Overholser asked, 'if the number of women in high office better reflected their share of the population? Would they then govern differently, feeling more comfortable and secure as women?'

She turned to the example of Norway, where the prime minister – Gro Harlem Brundtland – was a woman (and, incidentally, a physician). Of seventeen cabinet ministers, seven were women. And the consequences were dramatic. One example: Despite huge spending cuts, the Norwegian government increased child-care subsidies and paid parental leave. In Norway's case, the answer to Overholser's question was: 'Probably, if enough came to power.'

In *The Sexual Brain*, I asked: 'Is it wise for males to confront each other across an arms-control negotiating table? Is global security enhanced in an atmosphere charged with testosterone?' Today, more women have had seats at that negotiating table – think of Madeleine Albright and Condoleezza Rice in the United States; Margaret Beckett, Britain's first female foreign secretary; Angela Merkel, the first female German chancellor. But the questions remain. Is it inherently bad that males indulge in *Schadenfreude* – or is a taste for revenge a valuable prerequisite for dispensing justice? Is it admirable that females have more empathy for a cheater in pain – or a regrettable sign of weakness? Kinder, maybe. Gentler, maybe. But is it the best way to run a country?

I am optimistic that the answer to questions like these will eventually emerge from the synergy of science and society. Our relatively recent ability to image activity inside the human brain is a giant step forward and will allow us to better understand individual differences. I view a science of empathy as a realistic prospect: Data will replace simplistic slogans (like *Men Are from Mars, Women Are from Venus*) as a basis for making social decisions. And I am optimistic that the social sophistication a larger number of female legislators can bring to the 110th U.S. Congress can help create 'a more perfect union, establish justice, insure domestic tranquility, provide for the common defense, promote the general welfare, and secure the blessings of liberty to ourselves and our posterity.'

The Zombie Concept of Identity

DAVID BERREBY

Science writer; author of *Us and Them: Understanding Your Tribal Mind*.

I'm optimistic about a notion that I call the zombie concept of identity. I think this once powerful idea is losing its grip on the thoughts of scientists, laypeople, and our political leaders.

The zombie concept of identity is the intuition that people

do things because of their membership in a collective identity or affiliation. It's a confusion that begins with a statistically valid idea – if you define your terms well, you can speak of 'American behavior' or 'Muslim behavior' or 'Italian behavior.' But this does not justify the assumption that all Americans or Muslims or Italians are bound to behave that way by virtue of their membership in the category (a category that, often, you created).

Science has shown us that people are not obedient zombies who do what identity ordains. The connections between identity and behavior are never that simple, because human beings have many overlapping identities ('American,' 'middle-aged,' 'Episcopalian,' 'Republican,' and 'soccer mom' can all be attached to one person in a single morning). It's what we're doing and with whom that seems to determine which of these identities comes to the fore at a given time.

Even as these insights became familiar in anthropology, social psychology, and other disciplines, they did not appear in everyday discourse about terrorism, immigration, social change, and so forth. That has had a number of bad effects. Leaders who think that people become terrorists because they're Muslims, or that new immigrants will be hostile to a society because they're immigrants, are making decisions that are bad for their nations, perhaps even unsafe.

Lately, though, I see abundant evidence that people are rejecting the zombie concept of identity. Pop culture is rich in stories and images that remind people of overlapping labels. More important, political rhetoric is giving way to realism about human psychology. One example: The anthropologist David Kilcullen was quoted in a recent *New Yorker* explaining that jihadists are pulled into terrorism by their circle of friends and their social networks – that 'there are elements in human psychological and social makeup that drive what's happening. This is human behavior in an Islamic setting. This is not "Islamic behavior."'

The significance here is that Kilcullen is not, currently, an academic but an official of the U.S. State Department. My hunch is that people in charge of fighting democracy's enemies will increasingly have to deal with identity as it is, even as popular culture is ever richer in reminders that people have a lot of overlapping affiliations.

So I have more than a little hope that in ten years' time, people who take seriously the zombie concept of identity will be looked upon as we today look upon people who believe in witchcraft. That will mean that popular discourse is closer to the conversations of scientists working in the field.

From that may come a feedback loop that will be good for science: As we become more comfortable with the idea that people have multiple identities whose management is a complex psychological phenomenon, there will be more research on the central questions: What makes a particular identity convincing? What makes it come to the fore in a given context? Those are the issues that link society, mind, and body, and when we have more competing theories about them, we'll be closer to understanding why people are the way they are.

That doesn't mean I think scientists, or anyone else, will stop saying things like 'He was raised Catholic' or 'She's a typical Jewish mother' to explain human behavior. Copernicus died almost five centuries ago, and we still exclaim, 'What a beautiful sunrise!' But in the same way that educated people today know that it's really the earth that moves relative to the sun, I think in a few years it will be widely recognized that identities are not our masters – that we will arrive at a better understanding of what collective identity means and how it is made.

Us/Them Dichotomies Will Become Far More Benign

ROBERT SAPOLSKY

John A. and Cynthia Fry Gunn Professor of Biological Sciences and professor of neurology and neurological sciences, Stanford University; author of *Monkeyluv : And Other Essays on Our Lives as Animals.*

A truly discouraging thing to me is how easily humans see the world as dichotomized between Us and Them. This is everywhere evident in social anthropology – for example, in all those cultures where the word for the members of that culture translates as 'people,' in contrast with the nonpeople living in the next valley.

As a neurobiologist, I'm particularly impressed with and discouraged by some recent findings: There's a part of the brain called the amygdala that has lots to do with fear and anxiety and aggression. Functional brain-imaging studies of humans show that the amygdala becomes metabolically active when we look at a scary face, even when it flashes by so quickly that we aren't consciously aware of it. And some recent work – solid, done by top people, independently replicated – suggests that the amygdala can also become activated when we see the face of someone of another race.

Yikes.

That's an upsetting finding. But on the heels of those studies are follow-ups showing that the picture is more complicated. The 'other skin color = activated amygdala = the Other' can be modified by experience – such as having grown up in a diverse world. The more diverse, the more likely the amygdala is to become activated. Or 'experience' can mean whether you are subtly biased to think about the person in the picture categorically or as an individual. If you're cued toward individuating, your amygdala doesn't light up.

Thus it seems quite plausible to me that we are hardwired toward making Us/Them distinctions and not being all that nice to Them. But what is anything *but* hardwired is who counts as an Us and who counts as a Them. We are easily manipulated into changing those categories.

So I'm optimistic that with the right sort of priorities and human engineering (whatever that phrase means), we can be biased toward making Us/Them dichotomies far more benign than they seem to be now – say, by making all of us collectively feel like an Us, with Them being the space aliens that we must be convinced may attack us some day. Or designating as Them the mean, intolerant people without compassion. But I'm certainly not optimistic that we'll soon have political, religious, or cultural leaders likely to move us in that direction. Thus, inly a low-rent version of optimism.

Multilingualism in Europe

GLORIA ORIGGI

Philosopher and researcher, Centre Nationale de la Recherche
Scientifique; author of *Text-E: Text in the Age of the Internet.*

I'm optimistic about Europe. On May 30, 2005, the day after the
French rejected the proposed European Constitution, I was traveling on the Thalys high-speed train from Paris to Brussels for a
committee meeting at the headquarters of the European
Commission. The train was full of people my age – the late thir-
ties – going to Brussels as experts in various domains to
participate in various E.U. projects. I started chatting with my
neighbors. The conversation was light, mainly about restaurants
and bars in Brussels or new exhibitions and movies. Most of the
people I spoke with came from more than one cultural background, with two or more nationalities in the family: father from
Germany, mother from Ireland, grew up in Rotterdam, say. All of
us were at least bilingual, many trilingual or more. I quickly realized that asking the opening question of ordinary train
encounters ('Where are you from?') had become patently obsolete. The reality was at odds with the newspapers' and politicians'
cliché of the typical E.U. officer as a gray, strait-laced, hideously
boring civil servant wasting time inventing useless bureaucratic
rules. My neighbors epitomized the deep cultural change now
taking place in Europe. A new generation has grown up, people
born more than a quarter of a century after the end of the Second
World War and now moving around Europe to study and work –
meeting, dating, marrying, and having children with people from
other European countries and doing so as a matter of course.

More and more European children are growing up multilingual. They are unlike immigrants born in one culture and having to grow up in another. They are unlike children growing up in a monolingual, monocultural family that happen to be located in a wider multicultural environment. For these children, cultural and linguistic diversity is not just a part of the society at large, it is a part of themselves, a novel kind of cultural identity. Multilingualism is becoming an existential condition in Europe, good news for a continent in which national identities have been so powerful and have caused so much tragedy and pain in the past.

Multilingualism is not only an existential condition, however – it also affects our cognitive life. Recent research in developmental psychology shows that bilingual children are quicker to develop an ability to understand the mental states of others. Most children under the age of four cannot recognize that our behavior is often based not on the way things are but on the beliefs we hold – true or false – about the way things are. Bilingual children, interestingly, succeed in what is known in developmental psychology as the false-belief task (distinguishing between the reality and what someone believes to be the case) several months earlier than do those who are monolingual. A likely interpretation of these findings is that bilingual children have a more fine-grained ability to understand their social environment and, in particular, a greater awareness that different people may represent reality in different ways. My bilingual six-year-old son makes mistakes in French and Italian but never confuses contexts in which it is more appropriate to use one language than the other.

I believe that European multilingualism will help produce a new generation of children who are cognitively more flexible, who will have integrated their mixed cultural background into their own identity and cognition. It will become impossible for

European educational institutions to inflict upon such students their local 'sacred' values based on a 'higher civilization' – greater bravery, spiritual superiority, or what have you. They will have to update their educational programs for young people who recognize themselves neither in local foundational myths nor in a feel-good 'multiculturalism' predicated on the maintenance of sharply distinct cultural identities. This will help new generations to get rid of 'unreal loyalties' (to borrow Virginia Woolf's term) to nation, flag, or local customs and manners. Multilingual citizens of a European space will be more tolerant, less sensitive to local allegiances and partialities. Their tolerance of diverse cultural identities will be built from within, not learned as a social norm.

All this may be wishful thinking, projecting my own personal trajectory on the future of Europe. But I can't help thinking that being multilingual is the best and cheapest antidote to cultural intolerance, as well as a way of going beyond the empty label of multiculturalism by experiencing a plural culture from within. And, of course, this is not just an European issue.

We Have the Ability to Understand One Another

REBECCA GOLDSTEIN

Philosopher and novelist; author of *Betraying Spinoza*.

Sharp polarities between clashing points of view are wreaking all sorts of havoc in the world right now. Perhaps for many of us the divide that cuts closest to the quick is that between science, reason, and logic, on the one hand, and sectarianism, faith, and religion, on the other. My optimism is anchored to one aspect of human nature: We have the ability to understand one another. Evolution has bequeathed us a sketchy folk psychology, just as it has a sketchy folk physics. We come equipped with the understanding that we are engaged with others who manifest propositional attitudes: beliefs, desires, regrets, dreads, hopes – the whole gamut. We come equipped, too, with skills for discovering what those propositional attitudes of others might be.

Since at least the 1940s, social psychologists have been studying our capacity for attributing mental states to others. In one early important experiment (Heider & Simmel 1944), almost every single subject, when shown a short movie consisting of geometrical shapes moving on a screen, attributed propositional attitudes to the shapes. Subsequent research has strengthened the view that our capacity for mental attribution is universal and nearly reflexive – in short, an aspect of human nature.

Our folk physics – ideas about space and time, objects and forces – can be extended and deepened, refined and corrected, by that sophisticated enterprise we call science. So, too, can our primitive folk psychology be expanded and refined. We can even come to understand those whose propositional attitudes differ significantly from our own. We humans may never be able to know what it's like to be a bat, but Daniel Dennett could, in principle, know what it's like to be a believer (to hold that life has meaning only if it conforms to, say, some larger-than-life purpose) or to be the victim of a dread of death so overwhelming that comfort is gained only from denying the reality of mortality altogether. And so, too, Pope Benedict XVI could, in principle, understand the propositional attitudes of a proponent of naturalism determined to trim ontology to the entities required by science because of a higher-order desire never to be duped into believing something false, and therefore committed to the highest standards of empirical evidence. (Not all propositional-attitude-bearers share this higher-order desire not to be duped, which can come as a shock to many in the scientific community.)

Obviously, to understand the propositional attitudes of another is not to endorse them, nor is it the same as wanting them for one's own – although that can of course occasionally occur, as when we learn from one another. Still, to come to know better the propositional attitudes of others, grasping what the world is like for them, can be intrinsically interesting. It can also be useful – in fact, it is often essential to survival and reproduction. A seducer will get nowhere without at least a rudimentary grasp of the propositional attitudes of the seducee. It also widens the circle of sympathy promoting the dissemination of ethical attitudes.

And of course the most effective means of changing people's minds usually involves grasping the minds they already have.

Just as science improves our folk physics, we have an enterprise that extends the primitive skills of folk psychology, refining them into a way of achieving a complex knowledge of what it's like to have propositional attitudes and representational structures different from one's own. This enterprise is the narrative arts. What gives me any optimism at all in this dark season of dangerous divides is the trend among contemporary novelists to turn their attention to the divisive themes of the day. This effort to develop narrative techniques for taking the full human measure of such divides can only contribute to deepening our understanding of what lies behind irreconcilable differences – both apparent and actual. We will never become an attitudinally homogeneous species. Someone who desires above all not to be duped into believing something false will not be turned into someone who, say, desires above all to have his beliefs affirm his affinity with his community – or vice versa. However, it's instructive for both to make their way into the other's mind. There's even a slight chance that one or the other might be changed in the process (though the deepening of understanding isn't measured solely by conversions). So I'm tethering my optimism to the work of our contemporary novelists – which is probably another way of saying that I'm pretty darned pessimistic.

How Technology Is Saving the World

DIANE F. HALPERN

Professor, Claremont McKenna College; past president of the American Psychological Association; author of *Sex Differences in Cognitive Abilities*.

I was going to be modest about it, but the congratulatory messages have been slow to arrive, so I will have to toot my own horn. The exciting news is that I have been named *Time's* 2007 Person of the Year.

It is true that this is an honor that I share with every other user of technology, but I share well, and I hope you are as flattered as I am at this well-deserved recognition. In past years, *Time* has bestowed this honor on such luminaries as Bill and

Melinda Gates, George W. Bush, Rudolph Giuliani, and Ayatollah Khomeini, so we are in good company. All users of technology received this (unfortunately) nonmonetary award as an acknowledgment of the work we have done to change the world in dramatic and unexpected ways. It would be difficult to imagine a more diverse group of award winners.

I am writing this article from London, where the Queen just announced that her annual Christmas message – which, according to royal sources, is anticipated by millions (I did not independently verify this) – will be available via podcast for downloading onto iPods and other MP3 devices. In fact, all of her past Christmas messages are now available in multiple viewing formats. Although the Queen refused to respond to questions about whether she personally owned such a device, the younger royals were quick to say that they did. As this momentous news (just one example of the radical changes in how we communicate) illustrates, there is ample evidence that the Persons of the Year are changing the world.

In just the last year, bloggers destroyed some political careers (Mark Foley's, for example) and launched others (Barack Obama's). Technology has changed the way we get information, compute and pay our taxes, keep diaries, do homework, stay in touch, read books, learn almost anything, conduct research, make purchases, compose music, find mates, run businesses, diagnose diseases, and more. At the click of a mouse, I can listen to rap music from Poland or find an organ donor. I believe that technology will also profoundly change how we think about each other and that, for the most part, the change will be for the good.

There are many possible Doomsday scenarios in which technology depersonalizes our relationships and makes war more efficient and propaganda more believable. But I am a teacher, and teaching is, at its heart, an act of optimism. If you scrape the crusty veneer off even the curmudgeonliest of professors, you will find a core (sometimes it will be a small core) of optimism. Optimism is a way of viewing possible futures with the belief that you can affect things for the better.

A colleague recently lamented that there are no unexplored places left on Earth. As our daily lives have become increasingly international, there are also fewer unfamiliar foods to be tasted, exotic locations to visit, or social customs that are entirely foreign. Although most of us are still far too uninformed about the lives of people in other regions of the world or in other neighborhoods in our own city, it is also true that we now know more about one another than ever before. Like the other users of technology, I have come to know much more about the lives of people who are not 'like me' than I would have even half a generation ago. We are not close to obtaining global citizenship, but it is at least an idea that people can comprehend and debate.

Along with the billion or so other award winners, I can and do communicate with people all over the world at cable speed. Consider for example, the international gaming community that plays together on-line and at the same time shares its political, professional, and personal views. These people engage in a prolonged social intercourse that spans continents. Their heroes are the gaming experts, who come from many countries and backgrounds. The strangeness of 'foreigners,' which used to define the relationship between people of different religions, customs, races, and regions of the world, is disappearing as the rapidly increasing numbers of users of technology connect over time and space in ways available only to members of the same clan or village a few decades ago.

Social and political psychologists study in-group favoritism, which has been a root cause of most conflicts throughout time. The term refers to the tendency to believe that members of one's own group are more deserving than members of other, or out-, groups. Those undeserving others, whom we perceive as being more similar to one another than members of our group are (*they* all look and think alike), pose a threat to our group's right to whatever is scarce and valued – land, good jobs, clean water, and so on. Think of any conflict, past or present. Whether it is the Bloods and the Crips, Shiites and Sunnis, Tutsis and Hutus, African Americans and Hispanics, or the hundreds of years of the Crusades, these are all examples of 'us against them,' or in-group/out-group conflicts. Pessimists will point out that these conflicts have existed since the dawn of time and only a pie-eyed optimist could believe that intergroup conflict can be reduced by making 'the other' seem more like 'us.' But there is evidence that it can.

Psychologists in Northern Ireland have found that when people in divided societies come into contact in nonthreatening ways, they are in a better position to understand the other group's perspective and come to regard 'them' as equally human as members of their own group. This contact can lead to greater intergroup trust and even a measure of intergroup forgiveness for past wrongs. Similar approaches are being used by Turkish psychologists who believe that the Turkish-Armenian conflict has suffered from 'a lack of real relational space.' They are approaching this long-running conflict by finding ways that members of these two cultures can relate to one another, and the virtual world may be the safest place for these meetings. Psychologists in Israel and Palestine have been assisting teachers and students from both sides of that divide to write joint history books that incorporate narratives from the lives of all of the people who are living contemporary Israeli-Palestinian history. It is the ability to come together that forces each side to recognize the humanity of 'the other.' Although these projects are a mix of face-to-face and technology-assisted meetings, technology can be employed to expand these peace-waging efforts.

Technology is bringing people from diverse backgrounds together, with most of the meetings taking place in the virtual world. The challenge is to bring in those who are still excluded from the technology revolution, which includes the poor of the world and those whose governments are engaged in futile efforts to restrict access to information from the rest of the world. There will always be 'in' and 'out' groups, but these categories are becoming more fluid as we identify with a variety of different groups instead of defining ourselves by traditional demographic variables.

Allegiances now extend beyond national borders. I feel as distressed about the loss of the innocent lives of Iraqi citizens as I do about the loss of the innocent lives of the women and men in the U.S. military. I can view the suffering of each anytime, night or day, by logging onto the 'local' news in any part of the world. I can read the uncensored thoughts of anyone who wishes to share them on personal blogs and watch the home videos uploaded to YouTube and other public video sites. Government censorship is virtually impossible, and the ability to hear directly from ordinary people around the world has caused me to see our connectedness. We have only just begun to realize the profound ways that technology is altering our view of the 'other people' who share our planet. The use of technology to make the strange familiar will have an overall positive effect on how we think about others in our shrinking world. We are becoming more similar and connected in our basic humanness. And that is a good thing.

Neuroscience Will Improve Our Society

MARCO IACOBONI

Neuroscientist; director of the Transcranial Magnetic Stimulation Lab, UCLA.

Some time ago, I believed that a book could radically change society – an extreme form of optimism. One of the books I thought could change society was *Anti-Oedipus*, written some thirty years ago by the philosopher Gilles Deleuze and the psychiatrist Felix Guattari. At some point, I must have lost my belief in the power of books to change society. But the good news is that my belief is coming back. It is coming back in a slightly different form. What I am optimistic about is that neuroscience research will make our society a better one.

I have spent the last twenty years doing neuroscience research. To make a long story short, a concept that emerges from recent neuroscience research is that humans are wired for empathy. We have cells in our brains that make us understand one another in a simple, unmediated, automatic manner. But if our neurobiology makes us wired for empathy, why is our world so full of atrocities?

The explanation for this apparent paradox is probably as follows. The neurobiological mechanisms that make us wired for empathy work at a pre-reflective, automatic, implicit level. Our societies are built on deliberate, reflective, explicit discourse. The two levels, of implicit and explicit mental processes, rarely intersect; indeed, there is evidence that they often dissociate. This is probably why our religious and political belief systems, which operate at the deliberate, reflective level, are able to divide us in such a powerful way, even though our neurobiology should bring us together.

The good news is that awareness of neurobiological mechanisms that make us wired for empathy is entering the public discourse. This awareness won't go away; it will seep through the reflective level of our mental processes. Indeed, people seem to have an intuitive understanding of how neural mechanisms for empathy work. It seems that people 'recognize' how their brain works when they are told about it. Then they can finally articulate what they already 'knew' at a pre-reflective level. I'm optimistic that this explicit level of understanding of our empathic nature will at some point dissolve the belief systems that dominate our societies and that threaten to destroy us.

The End of -isms

MARC D. HAUSER

Psychologist and biologist, Harvard University; author of *Moral Minds: How Nature Designed Our Universal Sense of Right and Wrong.*

Racism, sexism, species-ism, age-ism, elitism, fundamentalism, atheism. These -isms, and others, have fueled hatred, inspired war, justified torture, divided countries, prevented education, increased disparities in wealth, and destroyed civilizations. For some, they represent ideas to die for; for others, they represent ideas to destroy. Though the targets differ, there is a single underlying cause: a brain that evolved an unconscious capacity for seeking differences between self and other and, once those differences were identified, seeking to demote the other for selfish ends. Such a capacity is like a heat-sensing missile, designed to seek and destroy. It achieves its ends by clever tactics that amount to an ever escalating arms race between demoting the other to the level of a pestilential parasite while raising the self (and its brethren) to the level of saints. This is the bad news.

The good news is that science is uncovering some of the details of this destructive capacity and may hold the key to applying a solution. My optimistic view? If we play our cards correctly, we may see the day when our instinctive prejudice toward *the other* will dissolve, when we gain greater respect for differences and, in the words of bioethicist Peter Singer, expand our moral circle. Here's the playbook, building on several recent ideas and scientific findings:

(1) *Decide what is fair by living behind a veil of ignorance.* The late political philosopher John Rawls argued that every human being will default to a selfish position – a bias that grows out of survival instincts and, in current evolutionary parlance, the biasing of genetic kin. To avoid such biases and achieve impartiality, we must imagine a set of principles that would apply while ignorant of others' political affiliations, wealth, age, gender, and religious background. As a device, the veil of ignorance works wonders, because it feeds into our selfishness. Let's say that I think university professors should be paid the highest salaries, while athletes should be paid the lowest. I can entertain this principle of income distribution only if I would be satisfied

in a world where I was a low-paid athlete. The veil of ignorance guarantees impartiality. Teach it to your children, friends, colleagues, and enemies.

(2) *Recognize the universality of our moral intuitions.* When you peel away the explicit rules of action handed down by such institutions as religion and government, you find a common moral code. Those with a religious background tend to believe that abortion is wrong and so too is euthanasia. Atheists tend to disagree. But remove the doctrinal rules and our intuitive moral psychology propels us – our species, that is – to decide what is morally right or wrong based on general principles concerning our own welfare and that of others.

(3) *Be vigilant regarding disgust.* Disgust is the most virulent of human emotions. Although it arose as an adaptive response to potential disease vectors – starkly, things that are normally inside but are now outside, such as vomit, blood, and feces – it is a mischievous emotion, sneaking into other problems, alighting, wreaking havoc on group structure, and then spreading. Throughout the history of warfare, every warring group has tagged their enemy with qualities reminiscent of disease, filth, and parasites. The imagery is overwhelming, designed to trigger the rallying cry. Though the destruction of 6 million Jews by the Nazis was made possible by an extraordinary advertising campaign, it was made all the more possible by the carefully crafted manipulation of disgust: In the Nazis' eyes, Jews were vermin, dirty, diseased, and thus disgusting. Wouldn't we all be better off without disgust?

What if we could remove this emotional card? Would we undercut our efforts to denigrate the other? There are some people who never experience disgust and don't recognize it in others, even though they experience and recognize all of the other familiar emotions – sadness, happiness, fear, surprise, anger. These people suffer from the genetic disorder called Huntington's chorea. Though they experience significant deterioration of the motor systems, they are disgust-free. So too are carriers who are presymptomatic. Although we don't know whether patients with Huntington's disease are immune to propaganda meant to engender prejudices in them, my hunch is that science will confirm this relationship. And if that is the case, perhaps modern molecular techniques will one day find a way to cure Huntington's but, along the way, work out a method to crank down or turn off our disgust response.

This is a playbook for today; it is not the ultimate answer. It provides, I believe, a ray of hope that someday we may see greater peace in this world, greater respect for the other.

The Globalization of Higher Education

JAMSHED BHARUCHA

Professor of psychology, provost, and senior vice president, Tufts University.

Having recently returned from a visit to universities in India with which Tufts has partnerships, I am optimistic about the future of higher education, in part because it is becoming more global.

National borders can no longer contain the most serious problems the world faces, be they economic, environmental, health-related, or political. Through education and research, universities play key roles in addressing these problems.

In order to take on these challenges, people must understand the world beyond their respective nations. This requires that universities provide curricular and travel opportunities to learn about other countries. It also requires that universities recruit a critical mass of students from abroad; the presence of international students contributes to the international education of all students, because learning from peers is as important as learning from a formal curriculum. I am optimistic that colleges and universities around the world will take these challenges seriously and respond in enterprising ways to optimize the world's intellectual capital.

The United States is at the short end of a global knowledge asymmetry: On average, college students in the U.S. have less knowledge about other nations and cultures than their counterparts have about the U.S. Our colleges and universities are acting to compensate for this asymmetry by strengthening curricular offerings and active learning experiences that are internationally focused.

Amid the discussion of the recent report of the secretary of education's Commission on the Future of Higher Education, little attention has been paid to its call for 'greater emphasis on international education.' The report correctly points out that 'the need to produce a globally literate citizenry is critical to the nation's continued success in the global economy.' However, higher education should equip us not only to seize the economic opportunities afforded by globalization but also to navigate an increasingly interconnected, crowded, and dangerous world. We fail to understand other cultures at our peril.

In partial recognition of this, in 2006 the Bush administration launched the National Security Language Initiative (NSLI) to 'dramatically increase the number of Americans learning critical need language skills,' focusing on Arabic, Chinese, Russian, Hindi, Farsi, and other central Asian languages. While NSLI is a welcome initiative, the funding is exclusively for language instruction, yet the ability to engage beyond national boundaries also requires cultural fluency. Cultural fluency involves knowledge of history, politics, religion, literature, and the arts. It involves knowledge of gesture, nuance, and context necessary to avoid misunderstanding. Fortunately, although the vision of NSLI is restricted to languages, colleges and universities are already creating more expansive programs for international learning.

For any nation, the recruiting of students from overseas evokes mixed feelings. As we seek to advance the globalization of higher education, we must dispel two myths about the influx of international students. One is the brain-drain myth, according to which the countries of origin are being robbed of talent. Take the case of the large numbers of graduate students recruited from India over the past three or so decades – mostly in science and engineering. The dire warnings about a brain drain have proved false. These expatriate Indians have helped fuel India's emerging economy by leveraging their American training and global experience. This group also has formed a bridge between India and the U.S. that is providing the two countries with new economic opportunities along with a stable political relationship. We are all better off when talent is realized to its fullest – even if it crosses borders. The global matching of talent with opportunity is not limited to science and engineering. The great American conservatories of music are filled with students of Japanese, Chinese, and Korean descent, as are the stages of our concert halls.

A second myth about the movement of students across borders is that the host country bears a net cost. If we again take the example of American universities recruiting Indian graduate students in science and engineering, the truth is that the host nation is getting a bargain. Arguably the most selective science talent search in the world is the entrance examination for the undergraduate programs at the Indian Institute of Technology. American graduate programs in science and engineering – as well as American industry – have long relied on this selection process and have skimmed off the top of the IIT graduating classes in order to meet the U.S. economy's demand for scientists and engineers. The IIT is funded by the Indian government, so we in the U.S. are cashing in on scientific talent selected and trained at the expense of the Indian taxpayer – who, in turn, gets a return on this investment.

A similar story can be told about medical education. In some of the best private medical schools in India, medical education is subsidized by clinical revenue (from patients). Graduates from these programs – recruited to the U.S. to meet a growing demand for doctors, postdocs in the life sciences, and other health professionals – have had their training subsidized by health care consumers in India.

In an interesting twist on globalization, some Indians are going to China to study medicine because of the shortage of medical school seats in India. Medical education (particularly clinical training) in the U.S. is becoming prohibitively expensive – for students, medical schools, and teaching hospitals – even as the demand for doctors and other health care professionals soars. Countries like India and China, with large numbers of patients and rapid growth in the hospital sector, are likely to become destinations for clinical training.

The American-born children of the Indian students who were recruited to graduate programs in the first wave several decades ago are now represented disproportionately in the student bodies of the top American colleges and universities. Experienced at negotiating between two cultures, this generation is contributing to the internationalization of the educational experience on campus. From Bollywood music to Bhangra dancing, our campuses are becoming incubators of cross-cultural knowledge.

Knowledge knows no national borders, and learning shouldn't either. Institutions of higher learning are taking the lead in reaching across nations to prepare global citizens and leaders for a world in which cultures are more interwoven than ever before.

The Power of Educated People to Make Important Innovations

NATHAN MYHRVOLD

CEO and managing director of Intellectual Ventures; former chief technology officer of Microsoft Corporation; physicist, paleontologist, photographer, chef.

It is interesting that pessimism seems to be the conventional wisdom – i.e., that the world is going to hell in a hand basket and things are getting worse. In the short run, pessimism is an easy bet. The news media, for example, would be a terrible business if there were only good news. Shocking bad news sells more newspapers (and generates more Neilsen ratings and Internet hits). Yet they need not worry about there being a dearth of bad news – it's only a matter of time before more comes in.

However, I think the focus on pessimism is hugely misleading. The pattern of the last five decades is that, by and large, the most important factors in human life have improved immensely. By and large, there is no better time to be alive than today, and by any rational estimate things will continue to improve.

Perhaps the biggest reason I am optimistic is that I am a great believer in the power of educated people to make important innovations. The trends in China and India and elsewhere toward educating millions of people in science, engineering, and technology is tremendously positive. It is trendy in some U.S.-centric circles to bemoan the fact that China and India are graduating more engineers than the United States is; indeed, the developing world has the potential to graduate more engineers than the U.S. has people. I view that eventuality with optimism, on the whole. There will be negative consequences, to be sure, and some naysayers will whine about them. But history is clear that the negatives of bringing high levels of education to heretofore undereducated people are more than outweighed by the positives.

And Now the Good News

BRIAN ENO

Artist, composer, record producer.

Sometimes we learn from our failures. The acceptance of the reality of global warming has, in the words of Sir Nicholas Stern in his report on climate change to the British government, shown us 'the greatest market failure the world has seen.'

The currency of conservatism for the last century has been that markets are smarter than governments, and this creed has reinforced the conservative resistance to anything resembling binding international agreements. The suggestion that global warming represents a failure of *the market* is therefore important. Technical solutions will, we hope, be found, but the process will need to be primed and stoked and enforced by legislation that will be regarded as big-government socialism in the present climate. The future may be a bit more like Sweden and a bit less like America.

If a single first instance of global governance proves successful, it will strengthen its appeal as a way of addressing other problems – such as weapons control, energy management, money laundering, conflict resolution, people-trafficking, slavery, and poverty. It will become increasingly difficult for countries to stay outside of future treaties like Kyoto – partly because of international pressure but increasingly because of pressure from their own populations.

Which brings me to my main reason for optimism: the ever accelerating empowerment of people. The world is on the move, communicating and connecting and coalescing into influential blocs that will move power away from national governments, with their short-range horizons, and out into more global, consensual groups. Something like real democracy (and a fair amount of interim chaos) could be approaching.

The Internet is catalyzing knowledge, innovation, and social change and, in manifestations such as Wikipedia, proving that there are other models of social and cultural evolution – that you don't need centralized, top-down control to produce intelligent results.

The bottom-up lesson of Darwinism, so difficult for previous generations to accept, comes more naturally to the current generation. There is a real revolution in thinking going on at all cultural levels: People comfortably cooperate to play games for which the rules have not yet been written, with people they've never met. They listen to music and look at art that is emergent, not predetermined, and they accept the wiki model of the open-source evolution of knowledge.

All these developments represent dramatic and promising changes in the way people are thinking about how things work, how things come into being, and how they evolve.

We Will Overcome Agnotology (the Cultural Production of Ignorance)

ANDRIAN KREYE

Feuilleton (Arts & Ideas) editor, *Suddeutsche Zeitung*, Munich.

Have you heard this one by Conan O'Brien? 'Yesterday a group of scientists warned that because of global warming, sea levels will rise so much that parts of New Jersey will be under water. The really bad news? Parts of New Jersey *won't* be under water.' Or this one by Jay Leno? 'Heating bills this winter are the highest they've been in five years, but the government has a plan to combat rising bills. It's called global warming.'

Not their best jokes, but last year global warming became one of the staples of late-night monologs.

This makes me optimistic, because jokes are hard evidence of sociological currents. A joke works on national TV only if most viewers understand the cultural reference in a split-second. If Leno makes fun of global warming – great! It's now part of the collective consciousness and a sign of general consensus.

This is quite a change from just three years ago. In the summer of 2004, Hollywood director Roland Emmerich released his disaster movie, *The Day After Tomorrow,* saying he wanted to use it to combat the widespread ignorance in the United States about climate change. His film had been written accordingly. To bring the audience up to speed, much of the early scenes had Dennis Quaid, as a paleoclimatologist, do a lot of reciting of scientific facts and fiction, not unlike the endless dialogs between Spock and Kirk that set up the outer-space realities in *Star Trek.*

Emmerich's movie triggered a chain reaction. Media attention to climate change rose. TV features about endangered polar bears created emotional impact. Sales of hybrid cars went up. Seven Northeastern states have signed the Kyoto Protocol, an initiative followed by more than 300 U.S. cities. Several Hollywood stars (Brad Pitt, Keanu Reeves, Leonardo DiCaprio) have worked on documentaries about global warming, which are raising the topic's profile for an audience normally uninterested in scientific matters. Now even notoriously fatalistic Christian fundamentalists see Earth as a gift from God that humankind has to protect.

This is of course not just about the power of pop culture. It's not even just about climate change. This is about a society's choice between listening to science or falling prey to what Stanford science historian Robert N. Proctor calls *agnotology* – the cultural production of ignorance. Such production has been booming: the editing of NASA reports about climate change, the political sanctification of coma patient Terri Schiavo, the introduction of intelligent design into curricula. All those efforts have created a widespread will to ignore facts and reason.

If a nation purposely kept in the dark about an imminent danger for so long manages to overcome public inertia and become acutely aware of a complex issue like global warming in the span of two years or so, it means that the power of reason is ultimately able to overcome the forces of ignorance driven by economic interests and religious dogma. This is a universally optimistic outlook on history. And do not forget that this ability to swiftly react as a collective is still most important in the United States. The number of leading research facilities, the economic power, the pioneering spirit, and the entrepreneurial verve put the United States in a leadership position that can affect not just global affairs but, down the line, maybe even the weather.

The Major Climate Makeover

WILLIAM CALVIN

Theoretical neurophysiologist, University of Washington, and affiliate professor of psychiatry and behavioral sciences at the School of Medicine; author of *A Brief History of the Mind: From Apes to Intellect and Beyond*.

Mention global warming at a social gathering and see what happens, now that skepticism has turned into concern and sorrow. They will assume you're a pessimist about our prospects. 'Not really,' I protest – which earns me a quizzical look.

'Wait a minute,' she says. 'If you're an optimist, why do you look so worried?'

'So you think it's easy, being an optimist?'

Many scientists look worried these days. We've had a steady diet of bad news coming from climate scientists and biologists. To become even a guarded optimist, you have to think hard. But the history of science and medicine shows that once you understand the mechanics of what's what, you can approach all sorts of seemingly unsolvable problems. I'm optimistic that we will learn how to stabilize climate.

Unfortunately, the window of opportunity is closing. Fifty

years have passed since the first unequivocal scientific warnings of an insulating blanket of CO_2 forming around the planet. Politicians apparently decided to wait until something big went wrong.

It has. We have already entered the period of consequences. Climate scientists have long been worried about their children's future. Now they are also worried about their own.

Our Faustian bargain over fossil fuels has come due. Dr. Faustus had twenty-four years of 'Party now, pay later' – and, indeed, it's exactly twenty-four years since Ronald Reagan axed the U.S. budget for exploring alternative fuels. This led to doubling our use of cheap coal, the worst of the fossil fuels. They're planning, business as usual, to redouble coal burning by 2030 – even though we can now see the high cost of low price.

The Devil's helpers may not have come to take us away, but killer heat waves have started, along with some major complications from global warming. We're already seeing droughts that just won't quit. Deserts keep expanding. Oceans keep acidifying. Greenland keeps melting. Dwindling resources keep triggering genocidal wars with neighbors (think Darfur). Extreme weather keeps trashing the place. All of this will get worse before it gets better.

More important, tipping points can lead to irreversible demolition derbies. If another big El Niño occurs and lasts twice as long as in 1983 or 1998, the profound drought could burn down the rain forests in Southeast Asia and the Amazon – and half of all species could go extinct within a year or two.

Time has become so short that we must turn around the CO_2 situation within a decade to avoid saddling our children with the irreversible consequences of a runaway warming. That means not waiting for a better deal on some post-Kyoto treaty.

It means immediately scaling up technologies we know will work, not waiting for something better that could take decades to debug. This isn't optional. It is something we simply have to do. The time for talk is past.

'I see why you're worried,' she says. 'But what's your optimistic scenario for dealing with this fossil-fuel fiasco?'

For starters, I think it's likely that the leaders of the major religious groups will soon come to see climate change as a serious failure of stewardship. And once they see our present fossil-fuel use as a deeply immoral imposition on other people and on unborn generations, their arguments will trump the talk-endlessly-to-buy-time business objections – just as such moral arguments did when we were ending slavery in the 19th century.

Second, the developed nations are fully capable of kick-starting our response to global warming with present technology – enough to achieve within ten years a substantial reduction in their own fossil-fuel use. How? Wind farmers will prosper, as pastures grow modern windmills to keep the cows company. Giant parking lots, already denuded of trees, are perfect places for acres of solar paneling (drivers will love the shaded parking spaces they create). The carbon tax will replace most of those deducted from paychecks and create a big wave of retrofitting homes and businesses. Big, brightly lit grocery stores with giant parking lots will compete poorly with warehouses that deliver Web and phone orders within the hour, like pizza. Neighborhood grocery stores will once again do a big walk-in business, and they will compete with the warehouses by offering 'green bicycle' delivery.

High-speed toll gates will become the norm on commuter highways. (Yes, I know, but remember that your paycheck was just enriched by eliminating income-tax withholding.) Speed limits will be lowered to 50 mph (80 kmh) for fuel efficiency, and (as in 1973) drivers will marvel at how smoothly the traffic flows. Double taxes will apply to vehicles with worse-than-average fossil-fuel consumption, reducing the number of oversize vehicles with poor streamlining. Hybrids and all-electric cars will begin to dominate new car sales.

A firm, fast schedule will be established for retiring or retrofitting existing coal-burning plants. My bet is that building nuclear power plants – France gets 78 percent of its electricity that way, New Jersey 52 percent – will prove safer, cheaper, and faster than fixing coal-burning plants.

On the quarter-century time scale, let us assume that the new rapid-transit systems will reduce car commuting by half. The transition to electric and hydrogen vehicles will shift transportation's energy demands to greener sources, including biofuels, geothermal, tidal, and wave generation. The highly efficient binding-energy extractors (BEEs, the fourth-generation nuclear power plants) will be running on the spent fuel of the earlier generations. The low-loss DC transmission lines will allow, via cables under the Bering Strait, solar-generated electricity to flow from the bright side to the dark side of the earth.

And in this twenty-five-year time frame, we ought to see some important new technology making a difference – not just improvements in what we already use. For example, we might encourage rapid adaptation of the whale's favorite food, the tiny phytoplankton, which provide half of the oxygen we breathe as they separate the C from the CO_2. Since the shell-forming plankton sink to the ocean bottom when they die, their carbon is taken out of circulation for millions of years. Forests can burn down, releasing their stored carbon in a week, but limestone is forever. If shell-forming plankton could thrive in warmer waters with some selective breeding or a genetic tweak, their numbers might double and start taking our excess CO_2 out of circulation.

But even if we invent – and debug – such things tomorrow, it can take several decades before invention makes a dent in our urgent problem. And all this assumes no bad surprises, such as the next supersize El Niño killing off the Amazon and, once we lack all those trees, increasing the rate of warming by half.

By mid-century, let us suppose that we have begun extracting more CO_2 from the atmosphere than we add. This will happen only if the technology of the developed world has become good enough to compensate for what's still going on in the over-stressed nations that are too disorganized to get their energy act together. When CO_2 levels fall enough to counter the delayed warming from past excesses, we will begin to see a reversal of droughts and violent weather – though the rise in sea level will likely continue, a reminder to future generations of our 20th-century Faustian bargain.

As Samuel Johnson said in 1777, 'When a man knows he is to be hanged in a fortnight, it concentrates his mind wonder-fully.' We need to turn on a dime – by which I mean close to what we saw happening in the United States after the bombing of Pearl Harbor:

> From a standing start in late 1941, the automakers con-verted – in a matter of months, not years – more than 1,000 automobile plants across thirty-one states . . . In one year, General Motors developed, tooled, and com-pletely built from scratch 1,000 Avenger and 1,000 Wildcat aircraft . . . GM also produced the amphibious

'duck' – a watertight steel hull enclosing a GM six-wheel, 2.5 ton truck that was adaptable to land or water. GM's duck was 'designed, tested, built, and off the line in ninety days.' . . . Ford turned out one B-24 every 63 minutes . . .
— Jack Doyle, *Taken for a Ride* (2000)

Now, there's a source of optimism: We've done it before. Indeed, GM currently needs a new purpose in life (and I'd suggest repurposing the manned space program as well). All of that talent is badly needed.

With great challenges come great opportunities, and I'm an optimist about our ability to respond with innovation. Countries that innovate early will have an economic edge over the laggards. Our present civilization is like a magnificent cathedral back before flying buttresses were retrofitted to stabilize the walls. Civilization now needs a retrofit for stabilizing its foundations. It will be a large undertaking, not unlike those that once went into building pyramids and cathedrals. I'm optimistic that the younger generation can create a better civilization during the major makeover – provided those currently in the leadership can stop this runaway coal train, real fast.

Climate change is a challenge to the scientists, but I suspect that the political leadership has the harder task, given how difficult it is to make people aware of what must be done and get them moving in time. It's going to be like herding cats, and the political leaders who can do it will be remembered as the same kind of geniuses who pulled off the American Revolution.

Optimism Has a Bright Future

TOR NØRRETRANDERS

Science writer, consultant, lecturer; author of *The Generous Man: How Helping Others Is the Sexiest Thing You Can Do*.

I am optimistic about optimism. It has a bright future.

Not long ago, a few decades back, it became fashionable to be pessimistic. Gloomy predictions were sexy and sure signs of a progressive mind. Any true intellectual was skeptical about the survival of humanity. There were good reasons for that. In the early 1980s, we seemed to be a little stuck. A major nuclear exchange between superpowers was a very real threat, the environmental crisis was scarcely recognized by the public (and much less by the politicians), and the world population seemed as though it would grow exponentially forever.

But the mood changed. A nuclear-disarmament spiral started, most populations in industrialized countries became aware of the environment and demanded political action, the growth of the world population started to slow down. The Internet and the Web seemed to offer means of change and democratization. During the 1990s, optimism boomed.

Boom! But then came the dot-com crash and 9/11. The liberating force of the Internet was still there, even if hopeful and greedy investors didn't make money on it; as we now know, it was released by the open-source movement and phenomena like Web 2.0. Despite terrorism, the safety situation has not really changed in any quantitative sense. Also, climate researchers now agree that the global warming is actually here. The explanation is our own stupid and wasteful use of fossilized deposits of energy – this in a world full of an energy flow from the sun that is much, much larger than we need.

So is it time to be pessimistic again? No way! The environmental problems are much too serious to be left to the pessimists. We need to make a change and do it right away. But we know what to do. We need a suite of high-tech solutions that will enable us to plug into the good old resource base of humanity: solar energy, recycled materials, and a decentralized, peer-to-peer network to enable information flow.

We are about to make a very old niche in nature habitable for a huge human population through the use of high-tech solutions from the info-, nano- and bio- toolboxes. We must awaken the enlightening spirit of reinventing everything, the future included. We have to rethink the civilization we have developed since the introduction of agriculture in the last great climate crisis, at the end of the last ice age. We have to create a Civilization 2.0, and we have to do it now. Fortunately we have already started creating the technology and the dreams. Human beings have a talent for crisis management.

Without optimism, we would not be here. The rational evidence for giving up is always there, but we have survived, thanks to the faith and the strength to go on despite all warnings. Optimism always wins, until nobody is around to know that it did not.

Optimize optimism!

Save the Arctic – Now

GREGORY BENFORD

Professor of physics, University of California, Irvine; author of *Deep Time*.

No one truly thinks we can slow global climate change within half a century – at least, no economist who has looked at the huge momentum of energy demand in the developing countries. Collapsing the fossil-fuel energy sources would send us all back to the 19th century.

So, despair? Not at all. Certainly we should accept the possibility that anthropogenic carbon emissions could trigger a climatic tipping point, such as interruption of the Gulf Stream in the Atlantic. But rather than simply urging an all-out effort to shrink the human atmospheric-carbon footprint, my collaborators and I propose relatively low-tech and low-expense experiments in changing the climate on purpose, instead of by mistake.

I am optimistic that if we understand climate well enough to predict that global warming will be a problem, then plausibly we also understand it well enough to address the problem by direct means. And we do not have time to waste.

The simplest way to stop warming is to reflect more sunlight back into space. This suggests a simple priority:

- Begin with regional, reversible experiments to define the science.
- Learn from these how well we understand our climate.
- Look for cooling effects.

- Focus on what we can do now, not what we can do eventually. We do not have 'eventually' – nature works at its own pace. Our grandchildren, and probably their grandchildren as well, will never see the CO_2 levels we had just last year.

Perhaps the simplest idea uses the suspension of tiny, micron-sized harmless particles, just enough of them to create a readily measurable shielding effect and at such altitudes that they will not rain out for some six months. A first test could be over the Arctic, since the warming there is considerable; the test might take place north of 70 degrees latitude, outside national boundaries. The atmospheric circulation patterns there will tend to confine the deployed particles, sweeping them around the pole but not far southward. And since few people live there, any side effects could be minimal.

The method seems clear:

- deploy the particles by airplane in the spring;
- measure the cooling below, using local sensors and space monitoring of the sea ice;
- detect whether the present retreat of sea ice toward the North Pole slows or even reverses. This will be a clear, visual signature than the region is cooling; and
- measure to see if ground temperatures will give more refined understanding. After the main heating during Arctic summer has passed, the particles will rain or snow out in fall, ending the experiment in predictable fashion.

This idea exploits our expanding understanding of the climate system. It also uses our historical knowledge of the marked cooling driven by sulfate aerosols sent into the high altitude air by volcanoes in the last several centuries. But sulfates interact chemically with the high altitude air. We can avoid that by using less chemically reactive particles, such as diatomaceous earth. Our aim should be to edit the incoming sunlight, not to interfere with our atmosphere's chemistry.

We can regard these ideas, and the scientific knowledge we gain from such experiments, as tools in a possible future technology. There could be many useful variables in such a climate technology, including particle size, particle nature, altitude deployed (and therefore duration in the atmosphere), and much else. We probably do not even know all the major influences we will find. If such an Arctic experiment works, cooling the region, it could tell us much about how to arrest Arctic warming and reverse the loss of sea ice.

Repeating this test over several years would advance the science. Public discussion could run in parallel, giving the sense that this momentous issue is being freely aired.

This idea is only the first step in making climate science, which has always been passive, into an active science. Astronomy was passive until the space program began to give us the power to explore the planets a half century ago. We now do experiments on the soil of Mars and in the atmosphere of Jupiter and Venus. With direct measurement comes a new era in any science. Passive observation and theory are never enough; experiment decides.

This is not a new transition in scientific style, but it is unique. We will live inside the experiment. We have far more at stake. That is why we should get started now and learn as we go.

Diagnosing global climate change is only the beginning. Restoring the stable climate we are losing is the long-range goal, but the science comes first. Saving the Arctic is the first, trial step. If we find that the pace of forced global climate change is unacceptably high, we could then put this idea to work globally, with all deliberate speed. There could be new side effects on the vastly larger global scale, and we would have to monitor the entire process very carefully. Some effects could be positive. Lessening solar UV radiation would lower skin cancer deaths, now about a million per year. Crops under less UV grow better and yield more food, especially in the tropics.

The main thrust is to carefully use our ability to attack warming at its roots – incoming sunlight now, carbon dioxide later. The climate system has great inertia and stirs slowly, but once altered it has a powerful momentum. It is a good idea to have methods like these on the shelf to deploy quickly, ready for use if the global environment worsens.

Given signals that the scarier scenarios of a warming climate might soon be upon us, we should act soon. Such preparations can also establish the political ground for widespread action. Humanity needs to get used to the idea of acting in this wholly new fashion, assuming our role as true stewards of the earth. Costs seem readily attainable – perhaps a few hundreds of millions of dollars for a multi-year Arctic experiment. High-altitude trials over the open ocean are little constrained by law or treaty, so show-stopper politics may be avoided. These first stages will be scientific experiments, not vast engineering projects.

A successful experiment could alter the terms of the global warming debate for the better. As economist Robert Samuelson recently said, 'The trouble with the global warming debate is that it has become a moral crusade when it's really an engineering problem. The inconvenient truth is that if we don't solve the engineering problem, we're helpless.'

The Return of Commercial Sailing Vessels

GEORGE DYSON

Science historian; author of *Project Orion: The True Story of the Atomic Spaceship.*

I am optimistic about the return of commercial sail. Hybrid sail/electric vessels will proliferate by harvesting energy from the wind.

Two near-inexhaustible energy sources – sunlight and the angular momentum of the rotating earth – combine, via the atmosphere, to produce the energy flux we know as wind. We have two well-proven methods of capturing this energy: windmills and sailing ships. Windmills are real-estate limited (since most available land surface is already spoken for) and distribution-limited (since wind-swept areas tend to be far from where large concentrations of people live). Sailing ships turn wind energy directly into long-distance transport, but the practice was abandoned in an era of cheap fuel. The prospects deserve a second look. It is possible not only to conserve but even to accumulate fuel reserves by sailing around the world.

Modern sailing vessel design, so far, has been constrained by two imperatives: the exigencies of racing (for sport or against commercial competition) and ability to sail upwind. Under favorable conditions, sails produce far more horse-power than is needed to drive a ship. At a marginal sacrifice in speed, this surplus energy can be stored for later use by running the auxiliary propulsion system in reverse. Such hybrid vessels will able to store large amounts of energy – in conventional batteries, in flywheels, or by disassociation of seawater – and would be free to roam the world.

The trade winds constitute an enormous engine waiting to be put to use. When oil becomes expensive enough, we will do so.

The Ozone Hole

STEPHEN H. SCHNEIDER

Climatologist, ecologist, Stanford University; author of *Laboratory Earth: The Planetary Gamble We Can't Afford to Lose.*

If you are a climate scientist, seeking optimism is like a scavenger hunt: You have to look in some strange places.

Consider the ozone hole that opened up twenty years ago and nearly instantly (in a few years, which is really fast in international-treaty terms) created positive action: the Montreal Protocol to ban ozone-depleting substances. That's a good example of optimism, my students often suggest – right? Well, maybe – because we knew about the likelihood of ozone depletion for fifteen years before the ozone hole was shown to be caused largely by human emissions. So, what has happened since the early 1970s? The chemical industries denied it, took out character-assassination ads in major media to cast doubt on the scientists doing the work, and hired attack-dog lobbyists to block action in Washington.

Where is the optimism in this little piece of scientific history?

There is some. Unbeknownst to most of us, the chemical industry actually believed the science was about right, preliminary as it was, and (despite their political posturing to the contrary) had been working on substitutes for ozone-depleting chlorofluorocarbons so that one day, when the scientific evidence and politics aligned for ozone action, they would be ready to don the green cloak – and, by the way, take over market share by selling the world their newly designed, greener, 'ozone-friendly' chemicals. So, was the ozone glass half empty or half full? One could legitimately see it either way: optimistically ('We acted when we had to') or pessimistically ('Why did it take a catastrophe in the making to get action we should have taken a decade before on likely science?').

A 'Climate Hole'?

Any lessons from the ozone affair for climate-change policy?

First, what would unambiguously signal the advent of global warming? Perhaps the 2003 European heat waves that killed some 40,000 well-heeled people? Hurricane Katrina and New Orleans, connected to recent evidence that global warming will statistically increase top-intensity tropical cyclones? Amazingly enough, despite all the pessimistic talk from environmental types about biogeophysical 'tipping points' in the climate system (e.g., a shutting down of the Gulf Stream, collapse of the Greenland ice sheet, and so on), the optimistic half of my brain thinks the only clearly demonstrable tipping phenomena are psychological/political: the symbolic events of climate impacts just mentioned and the popularity of Al Gore's recent documentary, combined with a big change in attitude by some large corporations that are no longer lobbying to prevent anti-global-warming measures. These include GE, BP, PG&E, Duke Energy, Wal-Mart, and many others.

These events actually *are* cause for optimism. But, lest the O. overtake the P. a bit too fast, what has been proposed as climate policy so far is only a palliative that will stop less than half the projected warming. The really dangerous events occur after a few degrees of warming, and we are still squarely on that pathway, even though we are making some optimistic sounds and a few small moves to change direction.

So, in the Optimism vs. Pessimism Derby, optimism has staged a comeback in the backstretch and is gaining on the more apocalyptic favorite. That is the good news. But without widespread acceptance of the need to reconfigure our energy systems – and put our overweight cars on a diet – the troubling favorite is still a few lengths ahead as we enter the homestretch. Scientists still need to keep explaining credibly the consequences of all actions – and inactions – so that we have a good chance to avert the highest climate-change risks that business-as-usual is steaming us toward. It is still doable to steer the ship toward safer waters, but opportunity diminishes the longer we delay.

Perhaps the most optimistic aspect for me is that the young scientists I work with understand the dilemmas and are dedicated to explaining the situation to all who will engage.

A New, Environmentally Sustainable Worldview

SCOTT D. SAMPSON

Associate professor of geology and geophysics, University of Utah; chief curator and curator of vertebrate paleontology, Utah Museum of Natural History; host of *Dinosaur Planet*.

Given the current array of critical environmental woes – global warming, habitat loss, and species extinctions, among others – one might assume that there is little room for optimism. Nevertheless, I am optimistic, albeit cautiously so, about a profound shift in human attitudes toward the environment.

The current view in the western world is a reductionist perspective that has been dominant for over 300 years. Founded by scientists such as Descartes, Newton, Galileo, and Bacon, reductionism regards the natural world as a series of machines, best understood by ever more detailed examination of constituent parts. This mechanistic approach has generated a plethora of scientific breakthroughs – quantum theory, genetics, high-speed computers, and the germ theory of disease, to name a few – with each intoxicating success fueling ever more intense investigation of nature's components. Yet it has also fostered a fundamental division between humans and the natural world, with the former envisioned as dominating the latter.

Moreover, the Cartesian perspective on nature has proved to have severe limitations within science. In particular, because of a myopic focus on the parts, little attention has been given to connections and relationships, let alone wholes. In response to this perceived gap in understanding, many disciplines have recently turned to a 'systems' approach that often unites once separate disciplines. Thus there has been a growing emphasis on interdisciplinary research, with, for example, geobiology and biocomplexity becoming legitimate fields of study. Simultaneously, many educators have begun to direct their efforts toward revealing the 'web of life,' including the myriad connections that link the living and nonliving aspects of nature.

The underlying themes of the outdated, mechanistic perspective are isolation and permanence, with objects perceived as relatively permanent and distinct from one another. In contrast, the new worldview celebrates the opposite concepts: connections and change. And once again, there is a firm grounding in science, which has demonstrated that natural systems are inextricably interconnected and continually undergoing change (particularly if one's perspective includes deep time).

Thanks in part to a global economy and the World Wide Web, the mantra of this new movement – 'Everything is connected' – has even made its way into the popular media. At a slow but increasing pace, people are becoming aware that their everyday decisions can have far-reaching, even global, effects. Surely there is hope and optimism to be found in the many recent movements toward sustainability, even if most of these efforts remain on a small scale.

Nevertheless, any optimism with regard to a growing environmental consciousness must be tempered with a double dose of reality. First, environmental changes are occurring at rates entirely unknown in human experience. To give just one case in point, the rate of species extinctions is about 1,000 times greater than has ever been typical in Earth's history. Indeed, the current human-induced mass extinction is on track to obliterate on the order of half of all species on Earth by the close of this century, with unpredictable (and perhaps unimaginable) ecological consequences.

Thus, we have little time to make this transformational leap. The next few decades will be pivotal.

Second, the transition to a sustainable worldview will not occur simply through a sufficiently heightened fear of environmental collapse. Such a fundamental shift will require no less than a transformation of our educational system – not only K-12 but higher education as well. We must equip parents and educators with the tools to be effective not only in communicating the science of natural systems but also in fostering passion for nature ('biophilia,' to use E. O. Wilson's term). This process will involve getting children outdoors early and often, so that they have a chance to forge bonds with nature. First and foremost, education should be aimed at teaching children and adults how to live well in the world. Ultimately, in order for this pressing venture to be successful, scientists must become directly involved, communicating science to a broad audience at unprecedented levels.

PCT Allows Individuals to Address a Global Problem

JAMES GEARY

Former Europe editor, *Time* magazine; author of *The World in a Phrase*.

I am optimistic that purchasing power can be brought to bear on the problem of climate change. For too long, climate change has been one of those huge, complex, difficult issues people felt they could safely ignore. The science was complicated and there were persuasive voices arguing that it was all a myth anyway. Plus, climate change was just too abstract to get worked up about. ('Sure, Bangladesh and the Netherlands might be aquatic theme parks in a hundred years, but yesterday my job was outsourced to India. Which problem is more urgent?') Now few people seriously dispute the science, and climate change is having tangible effects on our lifestyles. ('I may not care what happens to island nations and fragile ecosystems on the other side of the planet, but I'll be damned if my annual skiing vacation is going to be ruined for lack of snow.')

People are more informed and concerned, but they still feel there's not much they can do. A survey carried out in early 2006 by the U.K. Department of Environment, Food, and Rural Affairs found that 70 percent of respondents said they believed their lifestyles had an impact on climate change; only 40 percent of those people said they thought they could do something to change that. There are two basic motivations for human beings to alter their behavior: fear and self-interest. I'm not optimistic that that will ever be different; I am optimistic, however, that people are now genuinely scared by climate change and that an appeal can be made to their self-interest in the form of Personal Carbon Trading (PCT).

The European Union's Emission Trading Scheme, launched in 2005, has been a qualified success. The system allocates a set of emissions credits (representing the amount of carbon a firm is permitted to produce) to industrial installations. There is an effort under way to extend the scheme to the airline industry. If a company doesn't use all of its credits, it can sell them to other companies that have exceeded their quotas. The idea is to make it expensive for corporations to emit the greenhouse gases that cause climate change, while making it financially attractive for them to limit those emissions. Several organizations are now experimenting with applying this idea to individuals.

PCT would be the economic equivalent of microgeneration; instead of every person generating his or her own power by means of solar panels or wind turbines, each person would generate his or her own carbon portfolio. PCT would work on the same principles as the EU system, but individuals instead of industries would be allotted carbon credits, which they could sell, trade, or use as they see fit. PCT could also have potential social welfare benefits: Children could be given carbon credits at birth, which would then (one hopes) increase in value, so that by the time they were adults they could cash some of them in to fund their education, professional training, or a first home purchase. PCT would allow people to take individual action to tackle a global problem, thus easing the existential angst so many of us feel when confronting something this immense. We could help save the planet and make money in the process! What's not optimistic about that?

The Challenge Presented by Peak Oil

BRIAN GOODWIN

Biologist, Schumacher College, Devon, U.K.; author of *How the Leopard Changed Its Spots*.

I am optimistic about our ability as a species to respond to the challenge presented by peak oil, the end of the cheap energy era that has lasted about 200 years, and to enter a new cultural phase in our evolution. There are several key developments that, despite the unprecedented challenge of this transition, encourage me to believe we can make it. These come not just from our scientific and technological insights into ways of resolving some deep problems in present cultural habits but also from shifts of perception that are occurring in cultural values.

The primary factor in scientific insight that is producing a major shift of awareness is the recognition that our dependence on cheap fossil fuel to satisfy our needs and desires has now entered the phase of disruption of the complex web of relationships on which the life of our planet depends. This shift has come from an understanding of how the release into the atmosphere of carbon dioxide from its buried condition in oil deposits is causing average temperatures to rise worldwide. Among the many consequences are disturbed weather patterns, caused by the excess energy being dissipated through increasingly destructive hurricanes, and the rise in sea level as the polar ice caps melt, threatening coastal habitation, in particular the cities. This awareness is leading both to global action (as in the Kyoto Protocol and various forms of carbon trading) and to local initiatives to shift our energy source from oil to renewables. There's no guarantee that we will survive this learning process. Every species throughout evolution has either made the hard choices in learning to live the path of sustainable relationships with others or has gone extinct. We face the same two alternatives. We are special in our own way (as is every species) but no different regarding this fundamental dichotomy of life or death.

A shift has also begun within the culture of science itself, where it is becoming clear why our separation of nature from culture has been a useful but dangerous assumption. Although this distinction was made in modern science in order to separate the objective from the subjective – that is, the reliable knowledge of nature from the idiosyncratic expression of human creativity – it has now exceeded its usefulness, encouraging us to see nature as a separate reality outside us that is ours to use for our own purposes. However, we *are* nature, and nature is culture: We are embedded in and reflective of the principles that govern the rest of reality, not separate as a result of such evolutionary gifts as consciousness and language. So we are all participants in the same evolutionary adventure. This insight came first in physics when quantum mechanics showed us that nature is holistic – not causally separable into independent, objective elements – and 'subjective' observers are contributors to this reality. And now in biology we are learning that it is not the genome that makes the organism but the networks of molecular elements in and between cells that selectively read and make sense of the information in the genes, creating organisms of specific form. The nature of this creative agency is what we are currently trying to understand. As I read the evidence, this is leading us to the realization that organisms use language as part of their creativity, as we do. Networking is also the principle of Gaia – the complex pattern of relationships between living organisms and the earth, the seas, and the atmosphere which results in the remarkable properties of our planet as a place fit for continually evolving life. We are not passengers on the planet but participants in its evolution.

Finally, what encourages me to believe that we have a chance of getting through the most difficult transition we have ever faced as a species is the proliferation of new technologies, along with experiments in trading and monetary systems, that could result in robust local communities that are self-sufficient and sustainable in energy, food production, and other human needs. The key here is, again, interrelatedness and networking. Whatever renewable, sustainable energy process is used (solar, wind, water, biofuels, or other; the combination will vary with geographic location and bioregion) will become the basis of a trading system that links the components of the community into a coherent, holistic pattern of relationships responsive to local conditions and responsible in its actions toward the natural world. These local communities will trade with one another but will preserve their distinctness, so that diversity is both inherent and valued, unlike the homogenization of current global relationships. Whatever population size emerges in such organic human networks will necessarily be within the carrying capacity of the bioregion that supports it. Life will be comfortable but not indulgent; a life of quality will emerge. Our ability to make this transition is evident in powerful expressions of public awareness, as in this insight from *A Book of Miracles*:

Our deepest fear is not that we are inadequate.
Our deepest fear is that we are powerful beyond measure.
It is not our darkness but our light that frightens us most.

We do indeed have the power (and are equipped) to make the transition, though it requires a fundamental shift in what motivates us – a shift from fear of nature to a deep sense of connection with her. This new, organic way of living, which combines science, technology, art, craft, and ritual in unified, coherent patterns of learning and doing and celebrating, has become a dream to be realized, because it is not only possible but has also become necessary.

Once and Future Optimism

SETH LLOYD

Quantum mechanical engineer, MIT; author of *Programming the Universe.*

I am optimistic about the past. It's looking better and better every day. A couple of hundred years from now, when the Greenland and Antarctic ice caps have melted and sea levels have risen by 200 feet, our genetically engineered descendants will be sitting by their ocean-front property in Nevada reminiscing about high old times in those now-submerged cities of New York, London, and Tokyo. From their perspective, the past is really going to look good.

I'm also optimistic about the future. It is well within our power as a species to avert the environmental catastrophe envisaged in the previous paragraph. Prudent investment in carbon-conserving technologies and economic strategies can postpone or prevent entirely the more extreme consequences of global warming. I am hopeful that policy makers will realize that relatively small sacrifices made voluntarily now can prevent much larger, involuntary sacrifices later.

Let's be realistic: We human beings are addicted to damaging ourselves and others. When one rationale for conflict loses force, we seek out a new one, no matter how trivial, for prolonging the strife. Nonetheless, we are capable of pulling back from the brink. During the cold war, the strategy pursued by the United States and the Soviet Union was officially called MAD (for Mutually Assured Destruction): Anyone who started a nuclear war was guaranteed to be annihilated themselves.

While risky in the long run – if the radar confuses a flock of geese with an incoming missile, we're all dead – the strategy worked long enough for our leaders to realize just how mad MAD was and begin to disarm. We are currently on the brink of a major environmental catastrophe, but there is still time to pull back.

Even if global warming does flood most of the world's major cities, human beings will survive and adapt. Just how they will adapt we can't predict, but they will. Technology got us into this mess in the first place, by providing the wherewithal for modern industrial society. I am optimistic that our descendants will develop technologies to cope with whatever mess we leave them. The technologies for survival into the 23rd century need not be high technologies: Simple, low technologies of water and fuel conservation will suffice. If we're careful with our basic resources, there should be enough left over to keep on playing video games.

We need not even leave the world a mess. The key to using resources wisely is to distribute them fairly. If only because the global distribution of resources such as money and energy is currently so skewed, I am no more than guardedly optimistic that our increasingly globalized society can make progress toward a world in which each human being has equal access to food, clean water, education, and political representation. This optimism is tempered by the acknowledgment that the world's 'haves' have little motivation to share the world's resources with its 'have nots.' We are unaccustomed to thinking of democracy as a technology, but that's what it is: a systematic arrangement of human beings into a social machine that functions better in many respects than the social machine of totalitarianism. The technology we currently require is not a more fuel-efficient

SUV but, rather, a political system that gives each human being on Earth a voice in policy.

Finally, I am *wildly* optimistic about the future of scientific ideas. Wherever I travel in the world – first, second, or third – I meet young scientists whose ideas blow me away. The Internet distributes cutting-edge scientific work much more widely and cheaply than ever before. As a result, the fundamental intellectual equality of human beings is asserting itself in a remarkable way: People are just as smart in the Andean hinterlands and the Hindu Kush as they are in London and Los Angeles, and they can now participate in scientific inquiry with far greater effectiveness than ever before. Human beings are humanity's greatest resource, and when all those humans start becoming scientists, watch out!

The Shifting Ratio of Benefit and Cost

COLIN BLAKEMORE

Chief executive, Medical Research Council; Waynflete Professor of Physiology, University of Oxford.

Despite the pall of gloom that hangs over the world, I'm hugely optimistic that some things will be better in 2007 than they were in 2006. Take a couple of things that are on the minds of many scientists: climate change and embryonic stem cells. In both cases, the imperative for action on the basis of scientific evidence is clear, and in both cases other forces have intervened to frustrate progress.

The obstacles to ameliorating the effects of climate change are short-sighted commercial interests and short-term political interests. Some businessmen still believe that their fortunes and those of their shareholders are best served by ignoring the doom-mongers and pumping out carbon dioxide to make money. A few politicians – one, in particular – have believed that their political standing and their place in history are favored by denying the growingly obvious. But the consequences of climate change are accruing nonlinearly. A point must come at which the impact of change will fall within the near-point of those refractory industrialists and politicians. When that happens, the rules will suddenly reverse. Business and politics will both be better served by response than by denial. I predict that the tipping point will come this year. Political skeptics will become passionate converts, eager to claim the historical credit for having recognized the inevitable. The burners will become preservers.

I should make it clear that what I am optimistic about is the likelihood of a change in attitude, not, alas, about the probability of rapid success in the monstrous task of reversing the effects of a century of profligacy. We are going to have to live with the consequences of our parents' actions, and our children will have to live with the consequences of ours. The issue is whether our children's children will inherit a world worth living in.

With regard to embryonic stem cells, the barriers to progress are not economic but moral. On the one hand, biomedical science offers the hope of cellular immortality – the prospect of repairing a damaged brain, heart, or pancreas, just as grazed skin or a cracked bone mends itself. On the other hand, a substantial cohort of politicians and religious leaders (more exactly,

Catholic and fundamentalist Protestant leaders), especially in the United States and some European countries, fiercely oppose what they see as the taking of life in the interests of other lives. Although the balance of arguments seems quite different from that for climate change, the crux of the problem, interestingly, is again the power of intuition over the cold rationality of science. I have heard a pro-life lobbyist describe the collection of stem cells from ten-day-old embryos, surplus to the requirements of in-vitro fertilization, as 'the evisceration of little babies.' Life, it is argued, begins at the moment of conception. Most scientists would surely argue that a pre-implantation embryo, smaller than the point of a needle, without a single nerve cell, let alone any viscera, cannot possibly be considered a person. Defining the starting point of life is not a matter of dogma but of social consensus. As my friend the Nobel laureate Eric Kandel has put it: 'Life begins when the kids are all through college and the dog dies.'

Given these absolutist arguments, why should I be optimistic about a change in attitude toward stem-cell research in 2007? Because morality is, for all but the most stubbornly impervious to practical evidence, a matter of utilitarian dialectic. Yesterday's moral outrage has a way of becoming today's necessary evil and tomorrow's common good. Just as with climate change, what will cause a swing of attitude is the turning point of a mathematical function – in this case, the shifting ratio of perceived benefit to theoretical cost.

Not long ago, a team of scientists from the Institute of Ophthalmology, the Institute of Child Health, and Moorfields Eye Hospital in London (supported, I'm delighted to say, by the Medical Research Council) reported that they had restored sight to considerably more than Three Blind Mice by transplanting into their eyes immature photoreceptor cells (midway between stem cells and fully formed rods and cones). Rats that have suffered strokes have been vastly improved by the transplantation of nerve-making cells into their brains. The first attempts will soon begin at repairing severed human spinal cords with the help of transplanted stem cells. The evidence of likely benefit is growing fast – no miracles yet, but a trickle of hope, which is likely to become a steady stream in 2007. I predict that the immorality of not helping the undeniably living sick will soon outweigh the good of protecting the never-to-be-born.

The Sunlight-Powered Future

ALUN ANDERSON

Senior consultant and former editor-in-chief and publishing director of *New Scientist*.

I'm optimistic about a pair of very big numbers. The first is 4.5 x 1020. That is the current world annual energy use, measured in joules. It is a truly huge number and not usually a cause for optimism, as 70 percent of that energy comes from burning fossil fuels.

Thankfully, the second number is even bigger: 3,000,000 x 1020 joules. That is the amount of clean, green energy that annually pours down on the earth totally free of charge. The sun is providing 7,000 times as much energy as we are using, which leaves plenty for developing China, India, and everywhere else.

How can we not be optimistic? We don't have a long-term energy problem. Our only worries are whether we can find smart ways to use that sunlight efficiently and whether we can move quickly enough from the energy systems in which we are entrenched to the ones we should be using. Given the perils of climate change and dependence on foreign energy, the motivation is there.

Can it be done? I'm lucky that as a writer I get to meet some of the world's brightest scientists, and I know that there are plenty of radical new ideas out there for a future in which sunlight is turned straight into the forms of energy we need. Here are just three of my favorites, out of scores of great ideas:

(1) *Reprogramming the genetic makeup of simple organisms so that they directly produce usable fuels – hydrogen, for example.* That will be much more efficient than today's fashionable bioethanol programs, because it will cut out all the energy wasted in growing and harvesting a crop and then converting its sugars into fuel.

(2) *Self-organizing polymer solar cells.* Silicon solar cells may be robust and efficient, but they are inevitably small and need a lot of energy to make. Self-organizing polymer cells could be ink-jetted onto plastics by the acre, creating dirt-cheap solar cells the size of billboards.

(3) *Artificial photosynthesis.* Nature uses a different trick to capture light energy, whipping high-energy electrons away from photopigments into a separate system in a few billionths of a second. We are getting much closer to understanding how it's done, and even how to use the same principles in totally different nanomaterials.

But what of the pessimists' view that we are just too entrenched in our current energy systems to change? There is a worldwide boom in investment in green technology already under way. And there are many transition technologies coming into operation that enable practice runs for more radical genome reprogramming and creation of new nanostructures. Although the consensus view is that the sunlight-powered future won't be taking over until 2050, I'd place an optimistic bet that one of the many smart ideas being researched now will turn out to be a winner much earlier.

The Coming Solar Power Boom

OLIVER MORTON

Chief news and features editor, *Nature*; author of *Mapping Mars*.

I am not, by default, optimistic; it is an attribute that I take on as a duty more than out of temperament. Left to myself, I do not look out at the world and see a hopeful place – and did not do so even when the geopolitical state we are in was not as dreadful as it is now. But I have been convinced over the years that an outlook that gives play to hopefulness is by and large a better tool with which to help improve the future than the alternative. You are more likely to find solutions if you believe they are there than not. The trick for those of us without the requisite sunny state of mind is to find excuses for optimism that make the duty feel less dutiful.

My current optimism is for solar energy. Put simply, the sun provides more energy to the earth in one hour than humanity makes use of in a year. Of the non-fossil-fuel energy sources, all the big players that aren't nuclear – biomass, hydroelectric, wind – are ultimately driven by the sun. I am optimistic that direct solar conversion – photovoltaic cells and their future analogs – will take its place among and then surpass those technologies a lot faster than most people outside the field currently imagine. I am hoping for at least a terawatt of solar by 2025, two if we're lucky, and dramatic cuts in carbon-dioxide emissions as a result.

The locus for this optimism is California. A history of generous and far-sighted subsidy has built up the silicon-based Japanese and German solar industries over the past decades. Something similar now looks to be happening on the West Coast, where newer technologies are poised to benefit. There couldn't be a better-suited place: California, and in particular the Bay Area, boasts a near unique concentration of world-class research universities and national laboratories, and a large number of people well versed in the solid-state trades who are ready and able to move from semiconductors dealing with information to those dealing with energy. It is also well endowed with venture capitalists, many of whom combine their desire to make money with an urge to change the world. They largely share a network-first, build-from-the-periphery, revolutionize-the-whole-shebang mindset, suited to (and shaped by) the development of the Internet – an attitude compatible with the idea of decentralized solar-power generation. The optimism I need to psych myself up to attain seems to come naturally to such people, in such places.

New materials and new material-processing techniques

should allow the cost of installed photovoltaic capacity to be halved in the next few years, and there is room for considerable further improvement after that. Whereas wind power, nuclear power, and dams will not become radically cheaper to install, solar-power capacity will. It will also become more flexible, both physically and metaphorically, with new applications on new surfaces, from windows to clothing. Some of these applications may well be gimmicky and unsustainable, but one of the great advantages of the coming solar-power boom is that it promises a wide range of technologies both to compete for the main prize – cheap domestic and light-industrial electricity in developed and developing countries – and to find or create new niches.

The boom will not just be a matter of lower-cost manufacture or better efficiency. Systemwide solutions need to be found – new ways of accommodating solar materials architecturally, new technologies for storing energy, smart approaches to the electric grid, new financial arrangements and instruments that will allow people to get the benefits of solar electricity without necessarily taking on the capital costs of installation themselves. The sort of imagination that makes such things happen is far from unique to California, but it is abundant there and can be put to use. The big breakthroughs may come in Germany or Texas or China, and they will certainly have to be used in China and India if they are to have the hoped-for effect on carbon emissions. But it is in California that we see the most striking collocation of public interest, political support, research capacity, technological exuberance, entrepreneurial flair, supportive business ethos, smart capital – and, crucially, sunshine.

The Sorcerer's Apprentice

GREGORY COCHRAN

Consultant, Adaptive Optics; adjunct professor of anthropology,
University of Utah.

'In the sweat of thy face shalt thou eat bread' – it has always been that way.

Most people have been slaves of necessity, while the few who weren't lived by exploiting those who were. Although mechanization has eased that burden in the advanced countries, it is still the case for the majority of the human race. Limited resources (mainly fossil fuels), and negative consequences of industrialization, such as global warming, have made some people question whether American living standards can ever be extended to most of the rest of the world. They're pessimists, and they're wrong.

Hardly anyone seems to realize it, but we're on the threshold of an era of unbelievable abundance. Within a generation – sooner, if we want it enough – we will be able to make a self-replicating machine, first seriously suggested by John von Neumann. Such a machine would absorb energy through solar cells, eat rock, and use the energy and minerals to make copies of itself. Its numbers would grow geometrically, and if we manage to design one with a reasonably short replication time – say, six months – we could have trillions working for humanity in another generation. You might compare this process to a single blue-green alga, which replicates over the summer until it covers the entire pond. But unlike algae, a self-replicating machine would be programmed and controlled by us. If it could make its own mechanical and electronic parts, it would also be able to make toasters, refrigerators, and Lamborghinis, as well as the electricity to power them. We could make the deserts bloom, put two cars in every pot, and end world poverty, while simultaneously fighting global warming. It's closer than you think, since the key technologies are already being developed for use in rapid prototyping and desktop manufacturing. Aristotle thought that slavery would end only when looms wove by themselves. We're almost there.

Right now the human race uses about 13 trillion watts per day. The solar cells required to produce that much power would take up less than 0.2 of 1 percent of Earth's land surface; remember that Earth intercepts more solar energy in an hour than the human race uses in a year. That's still a lot of solar-cell acreage, but it's affordable as long as they make themselves. We could put them in deserts – in fact, they'd all fit inside the Rub' al Khali (the Empty Quarter) of Saudi Arabia. As I understand it, we like depending on the Saudis for energy.

But there are better ways. Solar energy works best in space – sure, the weather is better, but also consider that the vast majority of the sun's energy misses Earth. In fact, only about 1 part in 2 billion warms this planet. Space-based self-replicating systems could harvest some of that lost sunlight – enough to make possible a lot of energy-expensive projects that are currently impractical. An interstellar probe is a bit beyond our means right now, and the same is true of terraforming Mars. That will change within our children's lifetimes.

I'm reminded of Mickey Mouse as the sorcerer's apprentice in *Fantasia*. He enchanted a broomstick to fetch water but didn't know how to stop it. When he split the broom with an axe, over and over, each of the pieces took up a pail – and before you know, he was in over his head. But where he saw a crisis, we see opportunity.

Science on the Agenda

ADAM BLY

Founder and editor-in-chief of *Seed*.

I am optimistic that science is recapturing the attention and imagination of world leaders. Witness, for example, the agendas of the World Economic Forum, the Clinton Global Initiative, or the African Union Summit. Science has made a well-timed transition from a topic of peripheral interest to the leaders of the world to one inextricably tied to issues of development, global health, innovation, competitiveness, and energy. Spurring markets, arts, and ideas, science is making its way into our halls of power with considerable momentum.

The critical challenge is for our understanding of science to keep up with our growing interest in science. Our new global science culture demands a new level of science literacy – for general populations and, indeed, for the leaders who govern them. What constitutes a 21st-century science-literate citizen is one of the most important questions we need to collectively address today.

It's no simple task to convince nations struggling with clear and present threats to think about the future, let alone take action. But among leaders in the developing world, science literacy is emerging as a primary focus. The argument goes as follows: Move away from dependence on short-term relief and toward the development of a long-term scientific infrastructure that generates its own solutions. This entails an investment in people who will shape their own sustainable science culture. January's African Union Summit in Addis Ababa focused almost exclusively on this very topic – this on the heels of a consensus by the continent's education ministers that science 'is the most important tool available for addressing challenges to development and poverty eradication, and participating in the global economy.' China, for the first time, has made raising science literacy an official part of its development strategy. It is worth noting that China's plan calls for science literacy to extend across demographics – from urban workers to rural communities to government officials – each for different reasons but all for a common goal. We have heard much about the potential for the West to generate intellectual return on investment from its aid to the developing world – new insight into diseases, for example. It is exciting to imagine how these cross-continental laboratories may pioneer new approaches to science literacy with global consequence.

Science solves problems, and this should be its consistent tagline in the developing world. In the developed world, science will spark more than solutions; it can initiate a renaissance.

In Europe, the Large Hadron Collider, the biggest science experiment of our time and the herald of a new era of Big Science, will soon go on-line, corralling the collective imagination of (at least) a continent. Tony Blair has reaffirmed that Britain's 'future prosperity rests more than ever before on the hard work and genius of our scientists.' And Germany's physical-chemist-turned-chancellor Angela Merkel has made science one of the priorities for Germany's EU presidency.

In 1969, Robert Wilson, then the director of Fermilab, testified before the U.S. Congress in support of his multimillion-dollar particle accelerator. 'It has only to do with the respect with which we regard one another, the dignity of men, our love of culture,' he said. 'It has to do with: Are we good painters, good sculptors, great poets? I mean all the things we really venerate in our country and are patriotic about. It has nothing to do directly with defending our country except to make it worth defending.'

It will take inspired, informed, and heroic leaders to drive our global science culture forward – toward the development of Africa, the emergence of a renaissance, or outcomes we have yet to imagine. After an all-too-long period in which science and scientists seemed to have lost their seat at the table, I am optimistic that we are about to witness a new era of science savvy.

We Will Embrace the Reality of Progress

KEVIN KELLY

Editor-at-large of *Wired*; author of *New Rules for the New Economy*.

I am optimistic about the only thing – by definition – that we can be optimistic about: the future. When I tally up the pluses and minuses at work in the world, I see progress. Tomorrow looks like it will be better than today. Not just progress for me but for everyone on the planet in aggregate and on average.

No sane person can ignore the heaps of ills on this planet. The ills of a deteriorating environment, of inequality, of war and poverty and ignorance – and the ills of body and soul of billions of the planet's inhabitants – are inescapable. Nor can any rational person ignore the steady stream of *new* ills bred by our inventions and activities, including ills generated by our well-intentioned attempts to heal old ills. The steady destruction of good things and people seems relentless. And it is.

But the steady stream of good things is relentless as well. Who can argue with the benefits of antibiotics – even though they are overprescribed? Of electricity, or woven cloth, or radio? The list of desirable things is endless. While some have their downsides, we acknowledge the goodness of these inventions by purchasing them in bulk. And to remedy currently perceived ills, we keep creating new good things.

Some of these new solutions are often worse than the problems they were supposed to solve, but on average and over time the new solutions slightly outweigh the new problems. As Rabbi Zalman Schacter-Shalomi once said, 'There is more good than evil in the world – but not by much.' Unexpectedly, 'not much' is all that's needed when you have the power of compound interest at work – which is what culture is. The world needs to be only 1 percent (or even one-tenth of 1 percent) better day by day to accumulate civilization. As long as we annually create 1 percent more than we annually destroy, we have progress. This differential is so small as to be almost imperceptible, particularly compared with the 49 percent of death and destruction that is in our face. Yet this tiny, slim, shy discrepancy generates progress. But is there really even 1 percent betterment? The only evidence we have of it is people's behavior. When we watch what people do, we see them inevitably heading toward the increasing options and possibilities offered by the future.

No one I know has yet found a way to live in the future. Maybe someday we'll invent inexpensive time machines that allow us to vacation a hundred years into the future. Right now, if we want to live in 'tomorrow' – that place which is just a little better than today – the best we can do is live in the most forward-looking city on Earth. Cities are where the future happens, where there are increased choices and possibilities. Every day, a million people move from the countryside into cities, a journey that is less a trip in space than in time. These migrants are really moving hundreds of years forward, relocating from medieval villages into 21st-century sprawling urban areas. The ills of the slums are highly visible and don't stop the arrivals. They keep coming – as we all do – for the greater number of freedoms and options, which they never had before. This is why we live where and the way we do – to have 1 percent more choices.

Moving back into the past has never been easier. Citizens in developing countries can merely walk back to their villages, where they can live with age-old traditions and limited choice. If they are eager enough, they can live without modern technology at all. Citizens in the developed world can buy a plane ticket and in less than a day can be settled in a hamlet in Nepal or Mali. If you care to relinquish the options of the present and adopt the limited choices of the past, you can live there for the rest of your life. Indeed, you can choose your time period. If you believe that the peak of existence was reached in Neolithic times, you can camp out in a clearing in the Amazon. If you think the golden age was in the 1890s, you can find a farm among the Amish. We have lots of opportunity to head into the past, but few people really want to live there. Rather, everywhere in the world, at all historical periods, in all cultures, people have stampeded by the billions into the future of 'slightly more options,' as fast as they can.

Why? Because the future is slightly better than the past. And tomorrow will be slightly better than today. And while people's behavior confirms the essential reality of progress, progress is not something we have been willing to admit to in public. I am optimistic that in the coming years we'll embrace the reality of progress.

Cities Cure Poverty

STEWART BRAND

Founder of the *Whole Earth Catalog*, cofounder of The Well; cofounder of Global Business Network; author of *How Buildings Learn*.

Cities have always been wealth creators. Cities have always been population sinks. This year, 2007, is the crossover point from a world predominantly rural to a world predominantly urban.

The rate of urbanization is currently about 1.3 million new city dwellers a week, 70 million a year, still apparently accelerating. The world was 3 percent urban in 1800, 14 percent urban in 1900, 50 percent urban this year, and is probably headed in the next few decades to around 80 percent urban, which has been the stabilization point for developed countries since the mid-20th century.

Almost all the rush to the cities is occurring in the developing world (though the countryside continues to empty out in developed nations). The developing world is where the greatest poverty is and where the highest birthrates have driven world population past 6.5 billion.

Hence my optimism. Cities cure poverty. Cities also drive birthrates down almost the instant people move to town. Women liberated by the move to a city drop their birthrate right on through the replacement rate of 2.1 children per woman. No one expected this, but that's how it worked out. As a result, there will be another billion or two people in the world total by mid-century, but then the total will head down – perhaps rapidly enough to be a problem, as it already is in Russia and Japan.

Poverty in the megacities (those with a population of over 10 million) and hypercities (over 20 million) of the developing world will be highly visible as the disaster it is. (It was worse out in the bush, only not as visible there. That's why people leave.) But the poor who were trapped in rural poverty create their own opportunity once they're in town, by establishing their own cities, the squatter cities, where 1 billion people now live. They recapitulate the creation of cities past by generating a seething informal economy in which everyone works. The dense slums, if they don't get bulldozed, eventually become part of the city proper and part of the formal economy. It takes decades.

Globalization and urbanization accentuate each other. Medical care that couldn't reach the villages can reach slum dwellers. The newly liberated women in the slums create and lead community-based organizations, some linked with national and global NGOs) to handle everything from child care to microfinance. If the city has some multinational corporations closely watched by do-gooders back home, their pay rates and work conditions will raise the standard throughout the city.

This sudden urbanization is a grassroots phenomenon, driven by the resourcefulness and ambition of billions of poor people who are busy getting out of poverty as fast as they can. Some nations help the process (China is exemplary), some hinder it (Zimbabwe is exemplary), none can stop it.

Proviso: If climate change shifts from gradual to abrupt during the next twenty years, that bad news will obliterate the good news I otherwise expect in the realms of global population shrinkage and economic growth.

New Children Will Be Born

ALISON GOPNIK

Professor of psychology, Institute of Cognitive Science, University of California, Berkeley; coauthor (with Andrew N. Meltzoff and Patricia K. Kuhl) of *The Scientist in the Crib*.

New children will be born. This may seem rather mundane compared to some of the technological breakthroughs that other scientists have focused on; after all, children have been born for as long as the species has been around. But for human beings, children are linked to optimism in a way that runs deeper than just the biological continuation of the species.

Optimism isn't essentially a matter of the rational assessment of the future – it's an attitude rather than a judgment. And it's the most characteristically human attitude, the one that's built into our DNA. The greatest human evolutionary advantage is our innate ability to imagine better alternatives to the current world – possible universes that could exist in the future – and to figure out how to make them real. It's the ability we see in its earliest form in the fantastic pretend play of even the youngest children.

But in fact everything in the room I write in now – not only the computer and the electric light but also the right-angled wall and the ceramic cup and the woven cloth – was once imaginary, no more than an optimistic pipe dream. And I myself – a scientist, a writer, and a woman – could not have existed in the Pleistocene, or even in the only slightly less Neolithic atmosphere of the universities of fifty years ago.

This ability to change the physical and social world in unprecedented and unpredictable ways is deeply bound up with our characteristically extended human childhood, that long period of protected immaturity. The radical changes that have transformed human lives for the better never could have been accomplished in a single lifetime.

We change the world bit by bit, generation by generation. We pass on our own innovations and the new worlds they create to our children – who imagine new alternatives. Over generations, we can revise what we mean by leading a better life. Our moral lives are no more determined by our evolutionary past than our physical or social lives are.

I can only glimpse the future, and those glimpses are all heavily rooted in the past. But it's a good rational induction that my children and their children and all the new children to be born will see the world in new ways, discover new possibilities and find new ways to make them real, in ways that I can't imagine now.

A One-Way Ticket to Mars

PAUL C. W. DAVIES

Theoretical physicist, cosmologist, astrobiologist, Arizona State University; author of *The Cosmic Jackpot: Why Our Universe Is Just Right for Life*.

Sometime before the end of the century, there will be a human colony on Mars. It will happen when people finally wake up to the fact that two-way trips to the red planet are unnecessary. By cutting out the return journey, huge savings can be made and the way will then be open to establishing a permanent human presence on another world.

A one-way ticket to Mars is not an invitation to a suicide mission. Adequate supplies, including a nuclear power source, can be sent on ahead, and every two years more supplies, and more astronauts, will be dispatched to the new colony. Mars is relatively inhospitable, but it is far more congenial than outer space. It has all the raw materials needed for a colony to eventually become self-sufficient. To be sure, life would be cramped and uncomfortable for the trailblazers, but so it was for Antarctic explorers a century ago.

What about the risks of leaving people stranded on Mars?

Most of the danger of space flight is in the launches and landings, as the two shuttle disasters horrifically demonstrated. Eliminating the trip home would therefore halve the overall risk of accidents. The harsh Martian environment would undoubtedly reduce the life expectancy of the colonists, but astronauts on a round-trip would be exposed to comparable health hazards from months of space radiation and zero gravity.

Why would people go to Mars, never to return? There are many reasons: an innate sense of adventure and curiosity, the lure of being the first humans to open up an entirely new world, the desire to explore an exotic and unique environment, the expectation of fame and glory. For scientists there are added reasons. A geologist on Mars would be like a kid in a candy store and would soon chalk up a sensational publications record. The crowning achievement would be evidence for life, a discovery likely to transform our view of nature and our place in the cosmos. A straw poll among my colleagues convinces me there would be no lack of volunteers.

When might the first colonists set out? Within a few years, if politics doesn't stand in the way. NASA could send a crew of four to Mars with existing technology, but the agency lacks the nerve and imagination for such an adventurous mission. However, I am optimistic that the new players in space – China and India – will not suffer from Western timidity. A joint Indian-Chinese colony on Mars by 2100 is not only technologically feasible, it is also politically realistic.

Geomorphic Evidence for Early Life on Mars

GARNISS CURTIS

Geochronologist emeritus, University of California, Berkeley.

There are three geomorphic forms on Mars I have identified:

(1) the filling of old, deeply eroded impact craters with uneroded domes (in some cases, the eroded, original resurgent central domes are still visible);

(2) channels, hundreds of kilometers long, filled with similar dome-like material, which is arched in cross-section, the arches rising two or more kilometers above the channel margins; and

(3) vast areas of Mars, between latitudes 25° and 60°, covered with mat-like material approximately a kilometer thick in many places and completely burying Noachian-age impact craters as much as five kilometers in diameter.

The organic material composing these geomorphic forms is almost certainly cyanobacteria combined with calcium carbonate. Stromatolites on Earth, growing today and dating back to 3.4 billion years ago (though on a comparatively minuscule scale) are my model.

I believe that the growth of these gigantic stromatolitic forms on Mars gradually decreased the carbon dioxide in its atmosphere, leading to progressive cooling and finally to freezing conditions that ended the growth of the three geomorphic forms and the mostly warm, wet Noachian era. The cold, dry Hesperian and Amazonian eras that followed for the next 3+

billion years to the present have none of these geomorphic forms, proving that those forms needed water for their formation – and not just water but other essential materials.

Just as several types of meteorites from Mars have been identified here on Earth, it is likely that Mars has received meteorites from the bombardment of Earth. The question is, which planet seeded the other with life? We probably are not going to find the prokaryotic progenitors of the eukaryotic cyanobacteria here on Earth, but we have a good chance of finding those progenitors on Mars if they ever existed there, thus telling us which of the two planets life originated on.

By the Early 22nd Century, We Will Be Living on More Than One Little Tiny Ball in Our Solar System

RODNEY BROOKS

Director of the MIT Computer Science and Artificial Intelligence Laboratory (CSAIL); chief technical officer of iRobot Corporation; author of *Flesh and Machines: How Robots Will Change Us.*

I am optimistic about many things, especially the future. Not long ago, I met a number of people from the 22nd century, and they were delightful. We smiled and giggled together a lot, though none of them spoke a word of English. Even their Japanese was not so great just yet. But demographic analysis tell us that many of those little girls I saw in Kyoto will end up as citizens of the next century.

I am optimistic that even if none of the people I met do so, then at least someone who is already alive will be the first person to make his or her permanent home off Earth, and in the next century my new young acquaintances will go to sleep at night on Earth knowing that humankind has spread itself out into the solar system. Some people will have done it for wealth. Others, driven by our evolutionarily selected urges, will have done it to once again mediate risks across our gene pool by spreading out to different environmental niches. But the wonder of it all is that those now old but sprightly women in Kyoto will be able to revel in the romance of the human spirit, always questing to learn, understand, explore, and be.

Is this really going to happen in this century? Yes, really.

Government space programs in China, Europe, India, Japan, Russia, and the United States have all in recent months talked about their plans for the moon during the first quarter of this century. There is a new government-backed space race, less agitated than the last but more likely to produce sustainable technologies.

And then there are the billionaires and billionaire-lites. Richard Branson has teamed with aircraft-design maverick Burt Rutan (who won the Ansari X Prize with SpaceShipOne, funded by billionaire Paul Allen) to develop the world's first space airline, Virgin Galactic. They plan on putting 500 people per year into suborbital space. There are other suborbital and orbital space competitors (and collaborators), including Rocketplane Kistler, Space Adventures, and Benson Space Company, all driven by charismatic individuals. PayPal principal Elon Musk, through his company SpaceX, has developed the Falcon 1 and Falcon 9 vehicles and had his first launch, with a backlog of paying customers. Amazon founder Jeff Bezos is less public about his plans, but his company Blue Origins has been getting FAA licenses for low-altitude tests of a vertical take-off and landing system at his Texas ranch. And there is no shortage of other high-tech billionaires who have expressed interest in space ventures and may well be investing in some not yet announced. While all these efforts are building hardware for launches from Earth, serious discussions are under way about start-ups for companies that will provide higher-order services, including shipping asteroids to new orbits to deliver rocket fuel to paying customers. These are the first nanosteps of solar-system engineering toward the ultimate of a Dyson sphere of humankind's very own.

In the current activities, there are obvious analogies with heavier-than-air flight in the early 20th century, and look where that (with the help of a couple of world wars) got us in that century. There is no longer a mono- or duoculture for access to space and planetary bodies – certainly a reason to be optimistic about this second wave of space access. Even if this wave eventually sputters like the last, we still have plenty of time for a third wave during the lifetimes of my Kyoto friends. The key drivers will turn out to be either military or economic – or most likely both. Just as in 1907 the economics of heavier-than-air flight were not obvious, we are still struggling in 2007 with the economics of this new endeavor. But that, too, will come and will be the ultimate driver.

By the beginning of the 22nd century, humankind will have significantly raised the probability of its long-term survival by spreading its genetic material beyond Earth. That genetic material may be significantly modified from the current model – but that is another, and different, story. The point is that we will have spread ourselves to more than one tiny ball in our solar system and will continue to step to other systems and throughout the galaxy over subsequent centuries.

As for the coming events of this century, there may not be a Gion, but there will be bars on Mars, and over time they will gather their own histories and legends, as stories are told and retold. And just perhaps, one of my peek-a-boo playmates will be one of the great actors in the derring-do and swashbuckling courage under pressure that will surely be part of the coming adventures.

The Future of Human Mating

DAVID BUSS

Professor of psychology, University of Texas, Austin; author of *The Murderer Next Door: Why the Mind Is Designed to Kill.*

Each one of us has descended from a long and unbroken line of ancestors who mated successfully. They all found love, or at least a liaison. Evolution has forged a motivation to mate so powerful that it propels us to surmount daunting and demoralizing obstacles. The first problem is prioritizing conflicting preferences, compromising on some to attain others. Searchers must then sift through hundreds of options, limiting pursuit to potentials within shouting distance of attainability. Desirable mates bring out determined rivals, forcing fierce competition.

Complex and subtle attraction tactics must succeed in unlocking minds and melting hearts. Once past these minefields, there is no rest for the weary. After mating, sexual conflicts erupt, undermining months or years of effort. Mate poachers abound, threatening to lure away our lovers.

Infidelity diverts precious resources to interlopers and rips families apart.

Treachery runs rampant, spurned lovers rage, divorce rates rise. The modern world compounds these problems, from deception in Internet dating to cultural gaps revealed by cross-continental mating. Despite the obstacles both ancient and novel, I remain optimistic that humans in every generation will continue to succeed gloriously.

The Hedonic Set Point Can Be Raised

NANCY ETCOFF

Psychologist, Harvard Medical School and Harvard University's Mind/Brain/Behavior Initiative; author of *Survival of the Prettiest: The Science of Beauty.*

Most people are fairly happy, and the rest distribute into the very unhappy, fairly unhappy, or very happy categories. Wherever they fall on the happiness scale, people keep pursuing happiness. They want more. Can they get it? I offer a cautiously optimistic 'Yes.'

I'll admit that the initial yield from the new science of

happiness did little to support my optimism. It showed that happiness levels are durable, withstanding sweeping changes in health and wealth. Life changes, it suggests, but you don't. It showed that there is a substantial genetic component to happiness. People have a happiness baseline, influenced by personality traits (such as extroversion or neuroticism) that are partly heritable. The happiness baseline has been likened to the body-weight set point, leading some to believe that adding permanent points on the happiness scale is as likely as off-loading pounds permanently from the weight scale. You've got a battle on your hands, with a formidable opponent: yourself. Short of a biological fix, happiness interventions seemed doomed to formidable recidivism rates.

But the extreme picture of the human happiness baseline as a fixed set point and of adaptation to life events as inevitable and complete is wrong, and it's being revised rapidly While the 'average' person's happiness may bounce back to baseline, the happiness of many individuals does not. (About one in four people show a change of two or three points on a zero-to-ten scale, with 9 percent showing changes of three or more points; even stable people show an average of a one-point change in a recent study.) Personality is much less stable than body weight, and happiness levels are even less stable than personality.

I said 'cautiously optimistic' because, so far, for every person who shows a substantial lasting increase in happiness, there are two who show a decrease. Discarding the set-point idea for a more malleable happiness baseline means that we will uncover vulnerability as well as hope.

I am also optimistic that we will uncover diverse ways that people can find sustainable happiness. But we'll need to dig beneath the surface and resist one-size-fits-all formulas. I'll give one example. Some often-cited research suggests that married people are happier than single people (the never-married, the divorced, the widowed, the cohabitors). The latter group is large and getting larger. As of 2002, there were 86 million single adults in the United States; more than 40 percent of adults over 18 are single, up from 28 percent in 1970. Is this massive demographic shift dooming us to increasing unhappiness? Should we encourage people to marry to increase their happiness?

People marry for many reasons, but let's consider just their happiness. The newest research follows large groups of people over long periods of time. It finds that the average person adapts to marriage; after the first year or two, she is not any happier than she was before marriage (an alternate analysis of the data suggests that adaptation is incomplete but that happiness is increased by a tiny amount, 0.115 on a 0-10 scale). Looking beyond averages to individuals, the data show that some people return to their former happiness levels, some end up much happier, and about an equal number end up significantly less happy than they were before they were married.

Although this data will dismay some passionate advocates of marriage, it's good news both for the married and the single. The idea that marriage should make you permanently happier places a large burden on the already burdened state of marriage and creates unrealistic expectations in partners. But the fact that it can and does increase happiness for some, by a large amount and for a long time, is a thrilling possibility.

The data suggest that the demographic trends away from marriage do not portend an increasingly unhappy society. Along with other evidence, they suggest that what is important for happiness is the quality of a relationship, not its civil status.

Finally, forget optimism. I know this for sure: We will always form passionate bonds with others, and through them find joy, solace, comfort, love, amusement, sympathy, and moments of ecstasy, and we will know in them the awe and wonder of being alive.

Romantic Love

HELEN FISHER

Research professor, Department of Anthropology, Rutgers University; author of *Why We Love*.

'Love will find a way,' it has been said. But throughout most of our agrarian past, love did not, at least among the upper classes. I am optimistic about romantic love, because we are returning to patterns of romance that humankind enjoyed across most of our deep history: choosing lovers and spouses for ourselves.

Parents may have started arranging their children's marriages when the brain began to develop, some 2 million years ago. But in those few hunter-gatherer societies that still survive, parents initiate only the first wedding of a son or daughter. Moreover, this contract is flexible: If the callow newlyweds are not happy with their match, they pick up their few belongings and walk home. The contract has been honored, and the parents are pleased to see their child again. The young go on to choose their next partner for themselves.

But as our forebears began to settle down, some 10,000 years ago, and as they acquired immovable property, like fields of grain and sturdy homes, they began to need to cement their social ties. What better way than to wed your daughter to my son? Arranged marriages became a way to build one's fortune and secure one's genetic future. And these marriages had to endure. In some farming communities, you could fall in love with whomever you chose, but you married the 'right' individual, with the 'right' kin connections and the 'right' social, economic, and political ties.

The widespread tradition of arranged marriages began to dissipate with the beginning of the Industrial Revolution. As men and women left the farm for factory work, they no longer needed to maintain many of these connections. They could choose partners for themselves.

Today this movement is gaining speed, due to two dramatic world trends: the global rise of women in the paid labor force and the aging world population. For millions of years, women commuted to work to gather their fruits and vegetables and came home with much of the evening meal. Women were economically, sexually, and socially powerful. With the invention of the plow, women lost much of their economic independence. But as women go back to work and come home with money, they are reacquiring their economic autonomy – and their ancient ability to choose their lovers and spouses. With the aging world population, high divorce and remarriage rates, and many modern inventions, from Viagra to hip replacements, women (and men) now have the time, opportunity, and health to make their own match – what the Chinese call 'free love.'

And along with the rise of romantic love *within* marriage has come what sociologists hail as the 21st-century marital form known as peer marriages, symmetrical marriages, or companionate marriages: that is, weddings between equals. 'Marriage,' Voltaire wrote, 'is the only adventure open to the cowardly.' Today more and more men and women have the opportunity to enjoy this adventure – life with someone they passionately love. In this way, humanity is regaining a tradition that is highly compatible with our ancient human spirit.

Malthus Was Wrong

GEOFFREY CARR

Science editor, *The Economist.*

When I was growing up, the problem at the heart of every environmental question was human population growth: If there aren't many people around, what they do matters little; if there are lots, even careful living is likely to have bad environmental consequences. At that time, the earth's population was about 3 billion. It has now doubled to 6 billion. Not, on the face of things, great grounds for optimism.

The population curves in the newspapers and television programs of my youth went relentlessly upward. That was because they had only one exponential term. A real population curve, however, is logistic, not exponential. It does not rise indefinitely; eventually it reaches an inflection point and starts to level off. That is because a second term – in the form of lack of space, lack of resources, disease, or direct conflict between individuals – stabilizes it by matching the birth and death rates. And that was the fate the environmentalists of the 1970s predicted for humanity.

Such pessimism, however, failed to take account of the demographic shift that all populations (so far) have undergone as they have enriched themselves. The negative exponent is starting to show up. Its cause is not lack of space or resources, nor yet is it conflict or disease (even AIDS, malaria, and tuberculosis make only a small difference in global terms). Instead, it is what the doomsters feared most after population growth – economic growth.

As a quondam zoologist, I find the demographic transition in response to higher living standards hard to explain. On the face of things, better conditions should lead to larger families, not smaller ones. However, it is impossible to argue with the facts, and the facts are that the rate of population increase is dropping and that the drop is correlated with increases in personal economic well-being.

Perhaps the answer lies in the old idea of r- and K- selection. Indeed, the terms r and K come from variables in a two-term logistic equation that describes real population dynamics. K-selected species (which you may remember from your college ecology classes) have few offspring but nurture them lovingly. Those species that are r-selected have lots of offspring but display a devil-take-the-hindmost attitude to their issue's survival. The crucial point is that K-selected species live in safe, predictable environments, while r-selected species live in unsafe, unpredictable ones. If the individuals of a species were able to shift opportunistically between r and K strategies in response to shifts in the environment, then something like the demographic transition in response to wealth might be the result.

None of this means that the eventual human population of, say, 10 billion will be easy for the planet to support. But such support will not be impossible, particularly since economic growth in rich countries is less demanding of natural resources for each additional unit of output than is the case for growth in poor countries.

Malthus was wrong to observe that population increases geometrically while the resources available to support it increase arithmetically. It was an understandable mistake. It flies in the face of common sense that population growth will actually slow down in the face of better resources. But that is what happens, and it might yet save humanity from the fate predicted for it by the Club of Rome.

The Long View of Demographics

W. DANIEL HILLIS

Physicist, computer scientist; chairman of Applied Minds, Inc.; author of *The Pattern on the Stone*.

I am optimistic about humankind's ability to reach a sustainable balance with other life on Earth, in part because the number of humans on Earth will soon start to decrease. This doesn't mean that I think we should ignore our environmental problem – just the opposite. We should fight hard now, with the confidence that we can win a complete and lasting victory.

We are so accustomed to watching the explosion of human

growth and development that it is easy to imagine that this is normal. It is not. We are the first generation in history that has watched the human population double in our lifetime; no future generation is likely to see it again. All of those blights of growth we have come to accept – crowded cities, jammed roads, expanding suburbs, fish-depleted oceans, tree-stripped forests – are symptoms of a one-of-a-kind surge in human expansion. Soon they will be just memories.

There are currently over 6 billion people in the world. There will probably never be more than 10 billion. Population forecasts vary, but they all agree that human population growth is slowing. As people become more prosperous, they have smaller families. In every country where women are allowed free access to education and health care, rates of population growth are going down. Sometimes the trends are hidden by the delays of demographics, but the real population growth rates are already negative in Europe, China, and (if we subtract immigration) the United States. The total human population is still growing but not as fast as it once was. Assuming these trends continue, the total world population will be shrinking well before end of this century.

This long view of demographics allows me to be optimistic even though almost every other measure of environmental health is deteriorating. We are suffering from our binge of growth, and the parts of our world that are the last to binge are suffering the most. The binge is not just in size of population but also in the level of consumption. Yet here, too, there is reason for optimism. We are so wasteful in our use of resources that there are huge opportunities for improvement. With more efficient technologies, our fundamental requirements for food, materials, and energy should be well within the carrying capacity of our planet. We should be able to support the peak of human population at a higher standard of living than the richest nations enjoy today.

There is no doubt that the environmental challenges of the next decades are daunting, and they will require all the power of human striving and creativity to overcome. Yet I have no doubt that we will succeed. Innovation, goodwill, and determined effort will be enough to handle the next few billion people. Then, as populations shrink, demands on resources will be reduced. Nature will begin to repair itself, reclaiming what we have so hastily taken. I hope we manage to keep the gorillas, elephants, and rhinoceroses alive. By the end of the century, they will have room to roam.

Research Will Provide the First Effective Treatments for Many Diseases

IAN WILMUT

Director of the Centre for Regenerative Medicine, University of Edinburgh; coauthor (with Roger Highfield) of *After Dolly: The Uses and Misuses of Human Cloning.*

I am optimistic that, during this new century, research in biology and medicine will provide the first effective treatments for many diseases, although we cannot predict when they will become available; in some cases, it may take several decades.

A greater number of new treatments may well be developed

than were introduced in the last century. I make this judgment not only on the basis of a simple extrapolation from past developments but also in consideration of the new understanding and revolutionary techniques that are emerging: for example, the potential value of the genome-mapping projects, stem cells, and the techniques to assess many thousands of small molecules for their effects upon human cells in laboratory test systems. All of this is underpinned by rapidly advancing molecular biology providing essential understanding of the mechanisms that regulate cell function.

Entirely new opportunities arise from the mapping of the genomes of people, other mammals, and a variety of infectious agents that cause human diseases. Although we now know the entire genetic sequence of a small number of people and have new estimates of the number of genes in the human genome, we have a great deal to learn about the role of specific gene products and the mechanisms that ensure appropriate functioning of the genes. This stage in the development of human genetics will be far more demanding and will take far longer than mere mechanical reading of the sequence; however, in the end it will be rewarding.

We've known for some time that certain human diseases result directly from differences in DNA sequence, but despite considerable research efforts, only a small number of causative mutations have been identified. Modern rapid sequencing techniques will greatly facilitate these analyses. However, it is likely that in a far greater number of cases, sequence differences make people comparatively vulnerable to a disease but are not directly causative of it. These associations will be revealed only by large-scale studies, in which the genomes of hundreds, perhaps thousands, of people are determined while the incidence of diseases in that population is monitored. This may enable us to accurately warn people that they are vulnerable to specific diseases and to offer advice on lifestyle and medication to reduce that risk.

In time, information of this kind may also greatly increase the accuracy of selection of appropriate medication for particular patients. At present, because of differences in drug response in individuals, an adverse response to medicines is a major cause of death or the need for hospital treatment, even if the medicine is appropriately prescribed and taken. It is probably fanciful to think of tailoring medications for each person, because this implies a full knowledge of their likely response to and metabolism of every compound that might be prescribed. However, it's likely that understanding these mechanisms will lead to improved design and selection of new compounds.

A great deal has been made of the potential use of stem cells or their derivatives to replace those lost in degenerative diseases that reflect the death or malfunctioning of specific cell populations. Diseases considered suitable for such treatment include Parkinson's and other neurodegenerative diseases, juvenile diabetes, spinal-cord injury, and liver damage resulting from hepatitis or solvent abuse. In the haste to explore this use of stem cells, the potential benefit of using such cells for drug discovery and toxicology studies is overlooked. Drug assessment will be markedly more accurate as cells become available that are representative of the critical tissues of a variety of people.

In some cases, the cells will be genetically identical to those of patients with an overt inherited condition. There are a number of potential sources of such cells, but at present the most likely seem to be embryonic stem cells, because they are known to have two key characteristics. They can form all the various tissues of an adult, and they multiply almost indefinitely in the laboratory. In practice, this means that researchers will be able to study genetically identical cell populations over a period of years and examine their response to potential drugs. This is not true of any cells taken from adults.

The gene sequence known to be associated with a specific disease may be introduced into existing cell lines to create a population of cells that would be expected to exhibit the characteristics of the disease. Alternatively, it may be possible to use somatic-cell nuclear transfer from a patient with an inherited disease to obtain embryonic stem-cell lines having that characteristic, even if the causative mutation is not known.

In some cases, similar research may help us understand the molecular mechanisms regulating the function of stem cells in a tissue. In time, this may allow us to stimulate the replacement of damaged or lost cells from endogenous stem-cell populations in the patient. There are many practical advantages in this drug-based approach to cell therapy. The alternative is to produce cells of the required type from embryonic stem cells, in sufficient number that they can replace the lost cells. When they have reached the appropriate stage of their maturation, these must be inserted into damaged tissue in such a way that they can integrate fully into that tissue and restore normal function. While it is likely that each approach to cell therapy will be used for some diseases, there are clearly many potential benefits to a drug-based therapy.

I am optimistic that research has the potential to provide these new opportunities and many more not described; however, I am concerned that society tends to be frightened by innovations while taking for granted the treatments that are available. We will make the most rapid progress if we recognize that it was earlier research that led to the present treatments, and if we are excited by the challenges and opportunities arising from new research.

Early Cancer Detection

PHILIP CAMPBELL

Editor-in-chief of *Nature*.

Having always thought of myself as a perennial optimist, I was surprised at how challenging this *Edge* question turned out to be. I now realize that it's because my optimism is an attitude, rather than founded on careful estimation, and therefore bears little scrutiny. A corollary is that my optimism makes hardly any difference to what I manage to achieve in a typical day, apart from (importantly) getting out of bed in the morning.

Turning to Merriam Webster's dictionary, I confirm that optimism is either 'a doctrine that this world is the best possible world' or 'an inclination to put the most favorable construction upon actions and events or to anticipate the best possible outcome.' An attitude, therefore, of questionable robustness and idiotically dangerous in some circumstances. But among several similar definitions in the Oxford English Dictionary, I find something less loaded: 'Hopefulness and confidence about the future or the successful outcome of something.' This permits optimism to be rational, too.

And (to get serious) if there's one aspect of life wherein both rationality and hope are essential and seem to be paying off, it's in the battle against cancer. I focus my optimism on what currently looks like a peripheral flank in that battle but could – and I think eventually will – become a central focus of attention. We should of course be delighted by the few instances of drugs that hit a cancer target, even when the target wasn't the one originally intended. But just as important to me is the prospect of the use of proteins or other markers that permit the early detection and identification of cancer, hugely increasing the prospects of survival.

An early-detection cancer diagnostic needs to show low rates of false-positive and false-negative outcomes, should be able to distinguish tumors needing therapy from those that will do no harm, and should be acceptable in terms of cost and practicality. This combination is a tall order. But hope arises from the unprecedented sensitivities of mass spectrometers, single-molecule detection, and DNA amplification, not to mention the power of high-throughput biological screening. These bring us the almost unimaginable prospect of successful discrimination of cancer or – even better – pre-cancer marker molecules within the bloodstream. The recent discovery in mice of genetic pathways underlying progression from precursor to advanced stages of ovarian cancer is another milestone to sustain optimism.

Although the National Institute of Health has made early detection a priority, such work remains relatively underfunded in most cancer agencies. There are big challenges of clinical validation ahead. At the policy level, health planners and drug companies will need to be sure of its societal cost-effectiveness. These considerations, and the fact that diagnostics are less scientifically sexy than 'cures,' can deter researchers from pursuing early-detection studies. So it's precisely now that optimism needs to have bite, so that pioneering work in early cancer detection is championed and funded.

Cancer Stem Cells and Novel Cancer Therapies

STUART A. KAUFFMAN

Director of the Institute for Biocomplexity and Informatics, University of Calgary; author of *At Home in the Universe*.

In the past few years, evidence has increased that so-called cancer stem cells play a fundamental role in cancer. Typically comprising about 1 percent or less of a total tumor mass, these cells appear to have unlimited proliferation potential and the ability to drive cancer growth. Moreover, cancer stem cells have been implicated in metastasis. They have already been found in the leukemias, lung, colon, prostate, breast, skin, ovarian, and neural cancers. They may be present in all cancers. Their discovery may be the most important in cancer biology in the past half century. Cancer stem cells are likely to afford entirely new cancer therapies in the near future.

It becomes obvious that merely reducing the mass of a tumor without eliminating the cancer stem cells will almost surely lead to a recurrence of the disease. Thus, increasing numbers of investigators, including myself, are now focusing our efforts in three related directions:

(1) Finding the means to selectively kill cancer stem cells.
(2) Finding the means to stop cancer stem cells from proliferating.
(3) Finding the means to induce cancer stem cells to differentiate to nonmalignant cell types.

While it is foolish to regard cancer as simple, I believe that work on cancer-stem-cell therapy has a strong chance to dramatically improve cancer therapy within the next few decades. There are a number of approaches. For example, it is now possible – using a new discovery, siRNA – to 'turn off' the translation of the messenger RNA of a specific gene into its protein product. It is also possible, using other molecular-biology techniques, to overexpress any gene. These molecular techniques will allow investigators to try to perturb the activities of specific genes that control cancer-stem-cell behavior, in an attempt to attain the three aforementioned aims.

Further, high-throughput screening via robotics now allows small-molecule and other chemical libraries of high diversity to be screened to search for molecular perturbations that, if applied to cancer stem cells, achieve selective killing, cessation of proliferation, or differentiation to benign cell types. Our own laboratory and an increasing number of others are undertaking this work.

Differentiation therapy is already clinically effective in the case of treatment of acute myelogenous leukemia (AML) with vitamin A. The cancer cells differentiate into normal blood cells that do not proliferate. A research group recently screened a mere 1,700 chemicals and found eight that caused AML cells to differentiate to or toward normal nonproliferating blood cells. It is not improbable that by screening chemical libraries with thousands to hundreds of thousands of distinct compounds, molecules capable of selectively killing, shutting down proliferation, or inducing differentiation of cancer stem cells will shortly be found.

These approaches, however, must, as ever, be viewed with cautious optimism. For example, it may be that other cancer cells in a tumor can differentiate back into cancer stem cells. If so, they would require treatment, perhaps making cancer a chronic disease. The 'same' cancer from diverse patients may typically have accumulated different subsets of gene mutations, rendering moot the hope of finding a single magic bullet good for all cases of that cancer. Conversely, vitamin A is widely useful in AML, raising the hope that a modest number of compounds might treat most cases of the 'same' cancer in diverse patients.

Further, the relation between cancer stem cells and normal adult stem cells remains to be clarified. A treatment eliminating both cancer stem cells and normal stem cells of a given tissue could have untoward effects. Elimination of leukemic cancer stem cells might eliminate normal blood (hematopoetic) stem cells and affect the normal processes of these normal stem cells in blood formation. Conversely, one can hope that techniques will be found that can sustain the patient during therapy and regenerate normal blood cells from other stem cells in a patient, transplant them into the patient after cancer therapy, and overcome the normal stem-cell loss induced by therapy. And for some tissues – prostate, ovaries, uterus – loss of normal stem cells may not be grave.

The implications of cancer-stem-cell therapy are enormous, and the world scientific community is rapidly grasping its potential significance. It is important to stress that this effort will be 'big biology,' because techniques such as high-through-put screening and tests of patterns of gene activities using genetic microarrays are very expensive. Adequate funding will be required. Overall, I am deeply optimistic, as a doctor and a biological scientist, that we will at last find subtle ways to treat cancer either with stand-alone therapies or in conjunction with familiar surgery, radiation, and chemotherapy.

The Human Epigenome Project

JILL NEIMARK

Science journalist; coauthor (with Stephen Post) of *Why Good Things Happen to Good People*.

There are maps, and then there are maps. We're embarking on a kind of mapmaking that will usher in new ways of under-standing ourselves – a map that can explain why identical twins are not truly identical, so that one succumbs to schizophrenia while the other remains cognitively intact; why what your mom ate can save or sabotage your health (as well as that of your children and your children's children); and how our genetic fates can be tuned by such simple universals as love or vitamins.

It's the Human Epigenome Project (HEP), the next step after the Human Genome Project, which in itself was as audacious as the Apollo space program or the Manhattan Project, mapping 25,000 genes and the 3 billion pairs of bases in our DNA. And yet what the Human Genome Project mapped is like land without borders, roads without names, a map without movement. Genes are silent unless activated. To have them is not necessarily to be under their influence.

'Land lies in water, it is shadowed green,' begins Elizabeth Bishop's classic early poem 'The Map.' The double helix lies in the epigenome just as land lies in water. The epigenome is a flute playing a tune that charms the coiled snake that is the code of life, and the snake spirals upward in response. A long bundle of biochemical markers all along the genome, the epigenome responds to environmental signals and then switches genes off or on, up-regulates or down-regulates their activity. And in that change lies a great part of our destiny.

In 2003, in a widely discussed experiment, Randy Jirtle of Duke University Medical Center showed that he could change the activity of a mouse's genes by giving supplements to its mother prior to or during very early pregnancy. A mouse with yellow fur, whose offspring would normally also be yellow, will give birth to brown-furred babies if fed a diet supplemented with vitamin B12, folic acid, betaine, and choline. Even the offspring of the mother's offspring will be born with brown fur. The genes have not changed, but their expression has, and that change lasts for at least two generations. A fungicide used on fruit led to sperm abnormalities in rats, abnormalities passed down for at least four generations. This gives us insight into nature's ways: Apparently nature figures that any change in the food supply will last a while and isn't just a seasonal fling.

Moshe Szyf, Michael Meaney, and their colleagues at McGill University have shown that love can work in a similar way. If mothers don't lick, groom, and nurse their babies enough, a molecular tag known as a methyl group – a tiny molecule made of three hydrogen atoms bound to a single carbon atom – is added to a gene that helps regulate an animal's response to stress. In pups that aren't nurtured properly, the methyl group down-regulates the gene's activity for life. The pups have higher levels of stress hormones and are more afraid to explore new environments. What is nature saying? If a mother doesn't attend to her newborn, it's probably because the environment is hostile and stressful – so, better to be vigilant and cautious, even afraid. Later, Meaney and his colleagues showed that a common food supplement could do exactly the same thing to the genes of well-licked and nurtured rats. Once the pups were three months old, researchers injected a common amino acid, L-methionine, into their brains. This methylated the same gene, down-regulated it, and turned the rats into anxious wall-flowers.

In June 2006, the European Human Epigenome Project published its first findings on the methylation profiles, or epigenetics, of three chromosomes. The push to map the epigenome is on. I've recently seen various epigenetic stories coming across the science wires. From the University of Texas Medical Branch at Galveston came the news that breast-feeding protects children who are genetically susceptible to repeated ear infections because of common variants in their genes. The tendency toward ear infections runs in families, and researchers found the culprit in two gene variants that increase inflammatory-signaling molecules in the immune system. Remarkably, breast milk seemed to permanently quiet the genes, so that even later in childhood, long after the children had stopped breast-feeding, they were protected from recurrent infections.

In research from the Universidad Nacional Autonoma de Mexico and the Instituto Nacional de Cancerologia, Mexico, epigenetic drugs are now being studied in breast, ovarian, and cervical cancer. These drugs affect genes that, when reactivated, help regulate cell proliferation, cell death, cell differentiation, and drug resistance. They're cheaper than designer-name cancer drugs and might help increase survival rates.

Even water fleas are joining the epigenetic act. In a December study from the University of California at Berkeley, expression of genes in water fleas changed in response to common contaminants. Water fleas are regularly used to monitor fresh-water toxicity, usually with a 'kill 'em and count 'em' approach. Researchers found that copper, cadmium, and zinc decreased expression of genes involved in digestion and infection. Screening like this might help industry assess and avoid particularly toxic contaminants.

Epigenetics offers us a different kind of map, one where we can zoom in and zoom out – a map of many colors, with street signs, so we can navigate, with routes we can choose, destinations we can change. Maybe the gene isn't selfish after all; maybe it's actually sensitive. 'More delicate than the historian's are the mapmaker's colors.' So concludes Elizabeth Bishop's poem, and the epigenome may prove to be one of the more beautiful, delicate, and subtle maps of all time.

Growing Older

PETER SCHWARTZ

Futurist, business strategist; cofounder of Global Business Network; coauthor (with Peter Leyden and Joel Hyatt) of *The Long Boom: A Vision for the Coming Age of Prosperity.*

I am very optimistic about growing older. I turned sixty this year, and several decades ago I would have looked forward to a steady decline in all my physical and mental capabilities, leading into a long and messy death. The accelerating pace of biological and medical advances that are unfolding in front of us are heavily focused on reducing the infirmities of aging and curing the diseases of old age, or at least transforming them from fatal to chronic. There is a good chance that I will be among the vigorous new centenarians of mid-century, with most of my faculties working fairly well. Vision, hearing, memory, cognition, bone and muscle strength, skin tone, hair, and of course sexual vigor will all be remediable in the near future. Alzheimer's disease may prove to be curable, and most cancers are likely to be treatable and may be curable. Regenerative medicine may lead to a significant increase in youthfulness as new, custom-grown organs replace old, less functional ones. Within a few decades, we are likely to be able to slow aging enough to lead to lifetimes beyond a hundred and twenty years.

We Will Lead Healthy and Productive Lives Well Past Our Tenth Decade

LEO M. CHALUPA

Professor of neurobiology and ophthalmology, University of California, Davis.

I am optimistic that by the middle of this century it will not be uncommon for people to lead healthy and productive lives well past their tenth decade. This means that the high school kids of today who believe they will be forever young might well have their fantasy fulfilled, albeit in modified form. My optimism is based on three factors.

First, life spans in developed countries are getting progressively longer. So-called senior citizens are now engaging in activities previously reserved for those yet to reach what was once considered middle age. The current mantra, that today's sixties are the new forties, is more than just advertising hype. The reasons are complex, but certainly the psychological state of today's seniors – their refusal to simply accept old age – is a prime contributor.

The other two factors fueling my optimism stem from recent advances in biomedical sciences that offer not just hope but a virtual guarantee that we'll soon be living longer and better lives. Those recent advances come from two major research fronts.

There are some exciting results showing that manipulations of basic cellular functions can prolong longevity. The literature on this topic is too extensive to summarize here, but one example will suffice. A molecule called resveratrol (think red wine), produced by a variety of plants, has been found to significantly improve the life span of many different organisms by as much as 59 percent, and this even occurs in obese animals. The significance of the latter point is that until recently it was thought that the only way to increase longevity was by going on a strict starvation diet, but now it seems that you can eat your cake and extend your life span, too.

The other relevant scientific breakthroughs come from neurobiology, my field of expertise. We used to think that with age there was a progressive deterioration in brain-cell structure and function. But that widespread assumption has proved wrong. New nerve cells have been found to be generated in the brains of old animals, and we're learning more and more how this amazing property of the aged brain can be manipulated. Low levels of regular exercise, for instance, have been found to significantly enhance neurogenesis in the hippocampus, a brain structure that deals with memory. Moreover, a recent study from my laboratory showed that certain nerve cells in the eyes of old mice are capable of growing new processes. We have also found such growth of nerve cells in the eyes of old people. Then there is the tremendous promise of stem-cell research, which is still in its infancy, for replacing damaged or dysfunctional body organs.

Taken together, the implications of these and many other findings in the biomedical sciences are clear. We will be able to regenerate parts of the brain that have worn out or been damaged in the course of a lifetime, providing renewed capabilities to those who are currently considered old folks. So better start thinking what you'll be doing with all those extra years of life.

New Prospects of Immortality

MARVIN MINSKY

Computer scientist; first-generation artificial-intelligence pioneer, MIT; author of *The Emotion Machine: Commonsense Thinking, Artificial Intelligence, and the Future of the Human Mind*.

> *I wish it were possible . . . to invent a method of embalming drowned persons, in such a manner that they might be recalled to life at any period, however distant; for having a very ardent desire to see and observe the state of America a hundred years hence, I should prefer to an ordinary death, being immersed with a few friends in a cask of Madeira, until that time, then to be recalled to life by the solar warmth of my dear country! But . . . in all probability, we live in a century too little advanced, and too near the infancy of science, to see such an art brought in our time to its perfection.*
> – Benjamin Franklin to Jacques Dubourg, April 1773

Eternal life may come within our reach once we understand enough about how our knowledge and mental processes are embodied in our brains. For then we should be able to duplicate that information – and then to copy it into more robust machinery. This might be possible late in this century, in view of how much we are learning about how human brains work and the growth of computer capacities.

However, this could have been possible long ago, if the progress of science had not succumbed to the spread of monotheistic religions. As early as 250 BC, Archimedes was well on the way toward modern physics and calculus. So, in an alternate version of history (in which the pursuit of science did not decline), just a few more centuries could have allowed the likes of Newton, Maxwell, Gauss, and Pasteur to anticipate our present state of knowledge about physics, mathematics, and biology. Then perhaps by AD 300 we could have learned so much about the mechanics of minds that citizens could decide on the lengths of their lives.

I'm sure that not all scholars would agree that religion retarded the progress of science. However, the above scenario seems to suggest that Pascal was wrong when he concluded that only faith could offer salvation. For if science had not lost those millennia, we might already be able to transfer our minds into our machines. If so, then you could rightly complain that religions have deprived you of the option of having an afterlife!

Do we really want to lengthen our lives? (*Woody Allen: 'I don't want to achieve immortality through my work. I want to achieve it through not dying.'*) In discussing this prospect with various groups, I was surprised to find that the idea of extending one's lifetime to thousands of years was often seen as a dismal suggestion. The response to my several informal polls included such objections as these: 'Why would anyone want to live for a thousand hundred years? What if you outlived all your friends? What would you do with all that time? Wouldn't your life become terribly boring?'

What can one conclude from this? Perhaps some of those persons lived with a sense that they did not deserve to live so long. Perhaps others did not regard themselves as having worthy long-term goals. In any case, I find it worrisome that so many of our citizens are resigned to dying. A planetful of people who feel they don't have much to lose – surely this could be dangerous.

(I neglected to ask the religious ones why perpetual Heaven would be any less boring.)

However, my scientist friends showed few such concerns: 'There are countless things that I want to find out, and so many problems I want to solve, that I could use many centuries.' I'll grant that religious beliefs can bring mental relief and emotional peace – but I question whether these, alone, should be seen as commendable long-term goals.

What about the quality of extended lives? (*Anatole France: 'The average man, who does not know what to do with his life, wants another one which will last forever.'*) Certainly, immortality would seem unattractive if it meant endless infirmity, debility, and dependency upon others – but here we'll assume a state of perfect health. A somewhat sounder concern might be that the old ones should be dying in order to make room for young ones with newer ideas. However, this leaves out the likelihood that there are many important ideas that no human could reach in, say, less than a few hundred well-focused years. If so, then a limited life span might deprive us of great oceans of wisdom that no one can grasp.

In any case, such objections are short-sighted, because once we embody our minds in machines, we'll find ways to expand their capacities. You'll be able to edit your former mind or merge it with parts of other minds – or develop completely new ways to think. Furthermore, our future technologies will no longer constrain us to think at the crawling pace of real time. The events in our computers already proceed a million times faster than those in our brain. To such beings, a minute might seem as long as a human year.

How could we download a human mind? Today we are only beginning to understand the machinery of our human brains, but we already have many different theories about how those organs embody the processes we call our minds. We often hear arguments about which of those different theories are right – but those often are the wrong questions to ask, because we know that every brain has hundreds of specialized regions that work in various ways. I have suggested a dozen different ways in which our brains might represent our skill and memories. It could be many years before we know which structures and functions we'll need to reproduce. No such copies can be made today, so if you want immortality, your only present option is to have your brain preserved by a cryonics company.

Some have suggested that to make a working copy of a mind one might have to include many small details about the connections among all the cells of a brain; if so, it would require an immense amount of machinery to simulate all those cells' chemistry. However, I suspect we'll need far less than expected, because our nervous systems must have evolved to be insensitive to lower-level details; otherwise, our brains would rarely work.

Fortunately we won't need to solve all those problems at once. Long before we're able to make complete backups of our personalities, this field of research will produce a great flood of ideas for adding new features and accessories to our existing brains. Then this may lead, through smaller steps, to replacing all parts of our bodies and brains and thus repairing all the defects and flaws that currently make our lives so brief. And the more we learn about how our brains work, the more ways we will find to provide them with new abilities that never evolved in biology.

Personal Genomics

GEORGE CHURCH

Professor of genetics and director of the Lipper Center for
Computational Genetics, Harvard Medical School.

A small but crucial set of human pursuits has experienced
smooth exponential growth for many decades – sometimes so
smooth as to be hidden and then revealed with a jolt. These
growth industries involve information – reading and writing
complex artifacts made of electronic and/or DNA parts. The
iconic example is the personal computer, which, though traceable back to 1962, became manifest in 1993 when free Web
browsers spawned millions of personal and commercial Web
pages within one year. I'm optimistic that something similar is
happening to personal genomics this year. We are in free fall
from a stratospheric billion-dollar generic human genome
sequence (which only an expert could love) down to a sea-level
price for our own personal genomic data. Early adopters are
considering how to exploit it, surrounded by envious or obliv-
ious bystanders. We can now pinpoint the 1 percent of our
genomes that, in concert with our environment, influences the
traits that make us different from one another. Ways to tease out
that key 1 percent with 'next-generation' DNA-reading technology popping up, and to suddenly bring the street price down
to $3,000, are about as easy (or hard) to justify as buying some
bleeding-edge electronic gadget at an early stage when only
minimal software is ready.

I am optimistic that whereas society is not ready now, it will be
this year. The inevitable initial concerns about techno downsides

(for example, the Genetic Information Nondiscrimination Act of 2005) are already morphing into concerns about how to make these new gifts useful and reliable. Just this past August, the U.S. Senate began consideration of the Genomics and Personalized Medicine Act of 2006. Momentum is building for millions of people to volunteer to have their genome data correlated with their physical traits, to benefit the billions who will hang back due to inertia or uncertainty. These volunteers deserve up-to-the-minute education in genetics, media, and privacy issues. They deserve protection, encouragement, and maybe even rewards. Many current medical research studies do not encourage their human subjects to fully fathom the potential identifiability of their personal genome and physical-traits data or to learn enough to access and appreciate that data. The cost of educating the subjects is far less than the other costs of such studies and yields benefits far beyond the immediate need for fully informed consent. In contrast, other studies (like the Personal Genome Project) emphasize education sufficient to choose among (1) opting out of the study completely, (2) de-linking genomic and physical traits, (3) restricting linked data to qualified researchers, (4) allowing peer-to-peer sharing, or (5) a fully open public database. The subjects can redact specific items in their records at any point, realizing that items used to support conclusions in published work cannot be easily reversed. The excitement and dedication of these volunteers is already awesome.

I am optimistic that millions more will participate. Millions already do participate in old and new social phenomena, ranging from the Red Cross to Wikipedia, from MySpace/YouTube to SEC compensation disclosures. We wear ribbons of various colors and share personal experiences on topics that were once taboo, like depression, sexual orientation, and cancer. Rabbis' daily tasks now include genetic counseling. Our ability to track the spread of disease – not just HIV, bird flu, or anthrax but even the common cold – will benefit from the new technologies and the new openness, leading to a bio-weather map. We will learn so much more about ourselves and how we interact with our environment and our fellow humans. We will be able to connect with other people who share our traits. I am optimistic that we will not be dehumanized (continuing the legacy of feudalism and the Industrial Revolution), but we might be rehumanized, relieved of a few more ailments, better able to contemplate our place in the universe and transcend our brutal past.

Finding Mental Illness Genes

SAMUEL BARONDES

Neurobiologist and psychiatrist, University of California, San Francisco; author of *Better Than Prozac: Creating the Next Generation of Psychiatric Drugs.*

When I trained in psychiatry in the 1960s, schizophrenia and bipolar disorder were blamed on bad mothering. Now we know that the pathogenic stuff that mothers (and fathers) transmit is genes.

Once this was established, scientists began searching for the genes involved. The main thing we've learned so far is that variants of quite a few different genes (maybe dozens, or even more) may each increase the risk of developing one of these afflictions, and that several of these variants must work together in a particular person to produce an appreciable increase in risk. Because so many combinations of gene variants may each produce the same pattern of mental illness, it has proved difficult to identify any one of them with the techniques that identified the single genes that cause some other disorders, such as rare forms of Alzheimer's disease.

To have a good chance of identifying the combinations of genes that influence the development of schizophrenia or bipolar disorder, two things are needed: thousands of DNA samples from a person clearly suffering from the disorder under study (to compare with controls) and an affordable technique for scrutinizing each sample in sufficient detail to identify all the salient genetic variations. Both these requirements are now being met. Groups of researchers have been collecting the requisite number of DNA samples from patients with clear-cut cases of schizophrenia or bipolar disorder, and the costs of detailed genomic analyses keep coming down. When the numbers of samples are big enough and the costs of analysis are small enough, the relevant genes should be found. Already a few gene variants have been tentatively implicated as risk factors for schizophrenia.

Nevertheless, there are skeptics. Unconvinced by the tentative findings, their biggest worry is that there may be so many different kinds of schizophrenia or bipolar disorder that any collection of DNA samples, no matter how large, will be a jumble and that the many relevant gene variants in the samples will all elude detection.

This is where optimism comes in. I am optimistic that this approach can work now, if it is adequately funded. And I'm disappointed that there are enough influential pessimists to limit the wholehearted support required for success. Identification of these genes will have important practical consequences in the design of new treatments to replace the unsatisfactory ones we currently employ.

We're at the point where a concerted effort to find the gene variants that predispose to disabling mental illnesses has a high probability of success. It is a time for optimism. It is a time for funding with a full hand.

The End of the 'Natural'

ANDY CLARK

Professor of philosophy, University of Edinburgh; author of *Being There: Putting Brain, Body and World Together Again*.

I am optimistic that the human race will continue to find ways of enhancing its own modes of thought, reason, and feeling. As flexible adaptive agents, we are wide open to a surprising variety of transformative bodily and mental tricks and ploys, ranging from the use of software, sports regimes, and meditational practice to drug therapies, gene therapies, and direct brain-machine interfaces.

I am optimistic that, stimulated by this explosion of transformative opportunities, we will soon come to regard ourselves as constantly negotiable collections of resources, easily able to straddle the boundaries between biology and artifact. In this hybrid vision of our own humanity, I see increased potential not just for repair but for empowerment, expansion, recreation, and growth. For some, this same hybrid vision may raise the specter of coercion and subjugation. Clearly, not all change is for the better, and hybridization (however naturally it may come to us) is neutral rather than an intrinsic good. But there is cause for (cautious) optimism.

First, there is nothing new about human enhancement. Ever since the dawn of language and self-conscious thought, the human species has been engaged in a unique natural experiment in progressive niche construction. We engineer our own learning environments so as to create artificial developmental cocoons that have an effect on our acquired capacities of thought and reason. Those enhanced minds then design new cognitive niches that train new generations of minds, and so on, in an empowering spiral of coevolving complexity. The result is that, as artificial-intelligence pioneer Herbert Simon is reputed to have said, 'Most human intelligence is artificial intelligence anyway.' New and emerging technologies of human cognitive enhancement are just one more step along this ancient path.

Second, the biological brain is itself populated by a vast number of hidden 'zombie processes' that underpin the skills and capacities on which successful behavior depends. There are, for example, a plethora of such unconscious processes involved in activities from grasping an object to the flashes of insight that characterize much daily skillful problem-solving. To that standard mix, technology and drug-based enhancements add still more processes whose basic operating principles are not available for conscious inspection and control. The patient using a brain-computer interface to control a wheelchair will typically not know just how it all works or be able to reconfigure the interface or software at will. But in this respect, too, the new equipment is simply on a par with much of the old.

Finally, empirical science is at last beginning systematically to address the sources and wellsprings of human happiness and human flourishing, and the findings of these studies must themselves be taken as important data points for the design and marketing of (putative) technologies of enhancement.

In sum, I am optimistic that we will soon see the end of those overused, and mostly ad-hoc, appeals to the 'natural.'

A Breakthrough in Understanding Intelligence Is Around the Corner

TERRENCE SEJNOWSKI

Computational neuroscientist, Salk Institute; coauthor (with Patricia Churchland) of *The Computational Brain*.

The clinically depressed often have a more realistic view of their problems than those who are optimistic. Without a biological drive for optimism, it might be difficult to motivate humans to take on difficult problems and face long odds. What optimistic view of the future drives string theorists in physics working on theories that are probably hundreds of years ahead of their time? There is always the hope that a breakthrough is just around the corner.

In 1956, a small group of optimists met for a summer conference at Dartmouth, inspired by the recent invention of digital computers and breakthroughs in writing computer programs that could solve mathematical theorems and play games. Since mathematics was among the highest of human achievements, they thought that engineered intelligence was imminent. Last summer, fifty years later, another meeting was held at Dartmouth that brought together the founders of the field of artificial intelligence and a new generation of researchers. Despite all the evidence to the contrary, the pioneers from the first meeting were still optimistic and chided the younger generation for having given up the goal of achieving human-level intelligence.

Problems that seem easy – such as seeing, hearing, and moving about – are much more difficult to program than theorem-proving and chess. How could this be? It took hundreds of millions of years to evolve efficient ways for animals to find food, avoid danger, and interact with one another, but humans have been developing mathematics for only a few thousand years, probably using bits of our brains that were meant to do something altogether different. We vastly underestimated the complexity of our interactions with the world, because we are unaware of the immense computation our brains perform to make seeing objects and turning doorknobs seem effortless.

The early pioneers of AI sought logical descriptions that were black or white and geometric models with a few parameters, but the world is high-dimensional and comes in shades of gray. The new generation of researchers has made progress by focusing on specific problems in planning, computer vision, and other areas of AI. Intractable problems have yielded to probabilistic analysis of large databases using powerful statistical techniques. The first algorithms that could handle this complexity were neural networks with many thousands of parameters, which learned to categorize input patterns from labeled examples. New algorithms for machine learning have been discovered that can extract hidden statistical structure from large data sets without the need for any labels. Progress is accelerating now that the Internet provides truly large data sets of text and images. Computational linguists, for example, have adopted statistical algorithms for parsing sentences and language translation, having found transformational grammars too impoverished.

One of the most impressive learning systems is TD-Gammon, a computer program that taught itself to play backgammon at the championship level. Built by Gerald Tesauro at IBM Yorktown Heights, TD-Gammon started out with little more than the board position and the rules of the game, and the only feedback was who won. TD-Gammon solved the temporal credit-assignment problem: If, after a long string of choices, you win, how do you know which choices were responsible for the victory? Unlike rule-based game programs, TD-Gammon discovered, on its own, better ways to play positions and developed a surprisingly subtle sense of when to play safely and when to be aggressive. This captures some important aspects of human intelligence.

Neuroscientists have discovered that dopamine neurons, found in the brains of all vertebrates, are central to reward learning. The transient responses of dopamine neurons signal to the brain predictions for future reward, which are used to guide behavior and regulate synaptic plasticity. The dopamine responses have the same properties as the temporal difference learning algorithm used in TD-Gammon. Reinforcement learning was dismissed years ago as too weak a learner to handle the complexity of cognition. This belief needs to be reevaluated in the light of the successes of TD-Gammon and learning algorithms in other areas of AI.

What would a biological theory of intelligence look like, based on internal brain states derived from experimental studies rather than introspection? I am optimistic that we are finally on the right track – and that before too long, an unexpected breakthrough will occur.

AI Will Arise

JORDAN POLLACK

Computer scientist, director of the Dynamical & Evolutionary Machine Organization Laboratory, Brandeis University.

I often attack the original, 'we can program it' direction of the field of artificial intelligence, but I'm still optimistic that our primitive electromechanical and computing machines will one day become intelligent enough to be treated as living creatures. There won't be a way to prove they are intelligent or alive, but just as digitized music and rasterized print appear continuous because the differences are below our perceptual thresholds, electro-mechanical animations of intelligence will appear as alive because the differences will be less than we can detect.

I have a predictive sketch for how such intelligent machines might arrive. My definition of 'robot' is any device controlled by software and interacting with the physical world. An economically viable robot is such a system that earns a consistent return on investment (ROI). The ATM, the ink-jet printer, and the disk drive are today's omnipresent robots – they just don't seem like the robots of science fiction.

I see three streams coming together in the future to allow the emergence of lifelike humanoid robots. The first is the lowly Web-bot. Moving beyond scripted Eliza-like technology in chat rooms, these animated humans exist on Web pages, attracting customers, helping them navigate, or selling. While the first generation of companies seemed to have failed, currently Oddcast seems to be the most successful, while Haptek makes the most realistic Web-bots. These software puppets will enter a positive feedback loop as they are programmed to exploit human psychological weaknesses. When a virtual human becomes an effective sales machine for, say, life insurance or securities, it will have achieved the ROI necessary for further cycles of investment. (Another class of virtual humans are employed as extras in video games and movies, but the AI componentry has taken a backseat to dynamics and graphic realism.)

The second stream is animatronics. The word, coined by Disney, is about making mechatronic puppets that entertain us (Chuck E. Cheese, Big Mouth Billy Bass). But animatronics is fairly expensive and mainly used in big-budget Hollywood movies. The toy industry is capable of delivering inexpensive animatronics (Chatty Cathie, Teddy Ruxpin, Furby, Microsoft Barney, Robosapien, Pleo), but each one of these robot toys is a stand-alone design optimized for cheap manufacturing and mass marketing, rather than the result of an ever more capable animatronics practice. Nevertheless, I expect that some kind of animatronic toy platform, perhaps like the animated Chimp and Elvis heads from Wowwee, will eventually catch on and profits will drive efficiency until humanoid puppets are reasonably priced.

These two streams will then become symbiotic: The best salesbot software running on inexpensive humanoid animatronics can start replacing human salespeople in car showrooms and furniture stores. But these salesdroids will still be empty puppets, unless they are 'inhabited' by low-wage workers, over broadband.

The third stream has to identify and attack the core problem of AI – that sentient life forms are several orders of magnitude more complex than the most complicated systems designed and built by humans: our software. Building software using best practices always bogs down between 10 million and 100 million lines of code, at which point the software becomes unmanageable by teams of human engineers. Assume that a simulation of a sentient animal mind would take tens of billions of lines of code, just as bodies are made of tens of billions of living cells cooperating to form a whole. In order to understand how nature could design systems of far greater complexity than human engineers currently can, we must focus not on simulating human cognitive faculties or on trying to understand the brain but on the *process* that can design such minds and brains. Through work on evolutionary and coevolutionary machine learning, we have identified missing principles of evolutionary theory as implemented computationally. We've developed systems that surpass human designs of sorting networks and cellular automata rules, shown how coevolving robot bodies and brains could be automatically manufactured, and developed new incentive structures that can motivate a community of learners to become one another's teachers.

The third stream – sentience – I believe won't be programmed directly but will be a result of successfully replicating how evolution has achieved an open-ended self-organizing process on a computationally universal substrate. Once simulated sentience is achieved, it will happily reside, and earn a living selling used cars, in the aforementioned electronic puppets.

Technology in Education

DAVID DALRYMPLE

Computer scientist; researcher at MIT's Center for Bits and Atoms.

There's a lot in science and technology to be optimistic about, as evidenced by the numerous responses in this book to the *Edge* question, but I'll focus on the role of technology in education.

Before I entered college, I had never been enrolled in a school. Some of my education was provided by books, magazines, museums, and the like, but I feel the most useful was provided by technology. I was in the first generation to grow up with the Web as a fact of life, and I made use of on-line references and search engines every day to research topics in which I'd become interested. From early childhood, many of my questions were answered by a mix of university Web sites, ad-supported niche reference works, and charitable individuals sharing their own personal knowledge with the world. Today, Wikipedia alone provides peer-reviewed, freely contributed articles on over 1.5 million subjects, and Google indexed 25 billion items in 2005 (it no longer publishes the count). Almost piece of knowledge known to humankind can now be located on the Web.

New means of communication can also aid education. When I was seven, I e-mailed a science consultant whenever I had a question I couldn't find a ready answer for on the Web – questions such as 'Why don't the passengers in the Concorde hear a sonic boom?' and 'Where can I find the Bohr model of every chemical element?' In 1999, during the week of my eighth birthday, I used e-mail to contact the author of a book I really liked (*When Things Start to Think*), who happened to be Neil Gershenfeld, now my faculty advisor. I probably wouldn't have bothered to write a formal letter, so if e-mail hadn't existed then, my educational trajectory would have been entirely different. I was also mentored from many miles away by Ray Kurzweil, in a series of conversations enabled by e-mail; this was another major influence on my life.

Computing is a creative tool as well: It can be used to write essays (like this one), produce works of art (I've sold fractal-based art at local festivals), and write computer programs. Programming fascinated me from a very early age, but it wouldn't have kept my interest long if I hadn't had access to a computer. My experiences in programming may have been the most influential in my intellectual development: Problem-solving and critical thinking are rewarded, math skills are enforced, and I even wrote programs to help teach me things, like an arithmetic drill I wrote in LOGO at age five. I was also greatly aided throughout my college education in computer science by my earlier self-guided learning of many of the same concepts. Whereas I was taught eight programming languages in college, I've learned more than twice as many on my own, and those

my most valuable and (so far) useful learning

apert's constructionist theory best explains my
rience with education. In *The Children's Machine*
(1993), ─ ites: 'Constructionism is built on the assumption
that children will do best by finding ("fishing") for themselves
the specific knowledge they need. Organized or informal education can help most by making sure they are supported
morally, psychologically, materially, and intellectually in their
efforts.'

From this point of view, what holds back the education of
children in the developing world isn't so much a lack of school-
houses or qualified teachers but a lack of access to technology
and communications. Without the Internet, there's no good
place for these children to 'fish' for knowledge – the local elders
probably don't have a periodic table of elements on the hut
wall.

But I'm optimistic because the unstoppable force of Nicholas
Negroponte's charisma is now squarely facing off against this
problem. He's convinced a dream team of technical, educational, and political leaders to spend lots of money and time
working on it. His One Laptop Per Child (OLPC) project
shows no signs of failing, despite many reports to the contrary,
and it's moving at a breakneck pace toward a future not more
than a decade or two off, when every child in the world – developing and developed – really does have a laptop. Imagining
the possibilities is a start, but it seems that the OLPC team,
driven by the constructionist theories, has developed a host of
innovative hardware and software that promises to bring a
useful and creative education to the world.

I'm optimistic because my lab, the Center for Bits and

Atoms, with the aid of the National Science Foundation and other global organizations, is deploying 'Fab Labs' (Fabrication Laboratories) all around the world, from Boston to the Midwest to rural India and a village north of the Arctic Circle. Fab Labs bring something that even the developed world lacks broad access to: cheap and easy fabrication of physical objects and custom electronics. With a set of inventory, machines, and computers totaling roughly $50,000, those who enter the lab can make wooden furniture, high-gain antennas, and even $10-or-so 'thinner clients' – terminals that connect, over a variety of communications media, to $1,200-or-so servers that support hundreds of users. These types of objects can be and are developed by local inventors, or produced by oneself or in a 'micro-VC' community business, and they cost very little. Fab Labs are also another huge enabling factor for constructionist education (making things is one of the most useful and creative sorts of education).

Media artist Toshio Iwai, who was artist-in-residence at San Francisco's Exploratorium in 1991-92 and wrote the video game *Electroplankton*, tells a story about his mother taking away all his toys when he was a small child and telling him he could play only with toys he made himself. Iwai credits that moment as a turning point in his life – from passive to active, consumer to creator. I'm optimistic that, in the future, education will take place not in centralized houses of learning (places where students listen to lectures and then answer questions about them) but at construction sites, in art studios, in computing centers – places where useful and creative things are done. I'm optimistic that it will be a more useful and creative education and that it will produce people who are more useful and creative, who will contribute in turn to a more useful and creative society.

Science As a Broadly Participatory Activity

NEIL GERSHENFELD

Director of the Center for Bits and Atoms, MIT; author of *FAB: The Coming Revolution on Your Desktop – from Personal Computers to Personal Fabrication.*

I'm optimistic about the prospects for science becoming a much more broadly participatory activity rather than today's largely spectator sport.

Success as a scientist is certainly dependent on interest and ability, but it also requires access to the accumulated body of scientific knowledge and the means to practice it. I've found the former to be much more widely distributed than the latter. Until recently, membership in an elite technical institution was a requirement for becoming a successful scientist.

It's considered axiomatic that smart people like to surround themselves with other smart people, but the reality at a place like MIT is that we're all so time-stressed and multitasking that we rarely have time to do anything unscheduled; many of my closest collaborations are with people who are far away. Two technological changes now provide an opportunity to revisit the boundary between being on and off a campus.

The first is research on digital fabrication, which is leading to much more accessible means for making and measuring things. With $50,000 in infrastructure, a lab can now do experimental work in the field that would have been hard to do at MIT when I first arrived there. The second is the emergence of broadband video conferencing and software for project and knowledge management that can make remote collaborations as convenient as local ones. Together, these are leading to the emergence of technical training and research projects that are distributed rather than based on remote or centralized access to scarce resources. Instead of scientific careers being bounded by limited classroom and lab space, headcount numbers, editorial fashions, and overhead rates, they can be bounded by a much more interesting resource – the availability of ideas.

I expect that scientific productivity will always be nonuniformly distributed, with disproportionate contributions from a small number of remarkable people, but the sample size for finding and fostering those people can be improved by a few billion or so. There's a grand feedback loop ready to be closed between the content and the organization of scientific invention. Many of today's most compelling new questions are still tackled with old institutional models; it's ironic that religion has had its Reformation but that the role of a research university would be recognizable to a medieval monk. The future I'm optimistic about is one in which the creation, as well as consumption, of scientific knowledge is potentially accessible to anyone.

We Will Finally Get Mathematics Education Right

KEITH DEVLIN

Mathematician; executive director of the Center for the Study of Language and Information, Stanford; author of *The Math Instinct: Why You're a Mathematical Genius (Along with Lobsters, Birds, Cats, and Dogs.*

For the first time since Euclid started the mathematics education ball rolling over 2,000 years ago, we are within a generation of eradicating innumeracy and being able to bring out the mathematical ability that research has demonstrated conclusively is within (almost) everyone's reach. The key to this eventuality (actually two eventualities, one in the developing world, the other in affluent, technology-rich societies) is technology (actually two technologies).

First, the developing world. Forget the $100 laptop, which I think has garnered the support it has only because of the track record and charisma of its principal advocate, Nicholas Negroponte. The ubiquitous computing device that will soon be in every home on the planet is the mobile phone. Despite the obvious limitations of a small screen and minimal input capability, it will, with well-crafted instructional materials, provide the developing world with accessible education in the basic numerical and quantitative reasoning skills that will enable them to escape from the poverty trap by becoming economically self-sufficient. Such a limited delivery system would not work for an affluent consumer, who has choices, but for someone highly motivated by the basic desires of survival and betterment and who has no other choice, it will be life-transforming.

At the other end of the economic spectrum, the immersive, three-dimensional virtual environments developed by the gaming industry make it possible to provide basic mathematical education in a form that practically everyone can benefit from.

We have grown so accustomed to the fact that for over 2,000 years mathematics had to be communicated, learned, and carried out through written symbols that we may have lost sight of the fact that mathematics is no more about symbols than music is about musical notation. In both cases, specially developed, highly abstract, stylized notations enable us to capture on a page certain patterns of the mind. But in both cases, what is actually captured in symbols is a dreadfully meager representation of the real thing, meaningful only to those who master the arcane notation and are able to re-create from it the often profound beauty it represents. Never before in the history of mathematics have we had a technology that is ideally suited to representing and communicating basic mathematics. But now, with the development of manufactured, immersive, 3D environments, we do.

For sure, not all mathematics lends itself to this medium. But by good fortune (actually, it's not luck, but that would be too great a digression to explain), the medium will work, and work well, for the more basic mathematical life skills that are of the most value to people living in modern, developed societies.

Given the current cost of creating these digital environments, the revolution is some years away. We can also expect resistance from mathematics textbook publishers, who are making fortunes by selling a product that has demonstrably failed to work, and from school boards, who still think the universe was created by an old guy with a white beard 6,000 years ago. But as the enormous sales of video games drives their production costs down, the technology will soon come within reach of the educational world.

This is not about making the learning of mathematics 'fun.' Doing math will always be hard work, and not everyone will like it; its aficionados may remain a minority. But everyone will achieve a level of competency adequate for daily life.

Incidentally, I don't think I'm being swayed or seduced by the newest technology. Certainly I never thought that television or the computer or even artificial intelligence offered a path to effective math learning. What makes immersive 3D virtual environments the perfect medium for learning basic math skills is not that they are created digitally on computers. Nor is it that they are the medium of highly seductive video games. Rather, it is because they provide a means for simulating the real world we live in, out of which mathematics arises, and they do so in a way that brings out and confronts the player (i.e., learner) with the underlying mathematical structure of our world. If Euclid were alive today, this is how he would teach math.

The Nervous System of the Human Race Has Come Alive

ALEX (SANDY) PENTLAND

Director of MIT's Program in Developmental Entrepreneurship.

Ten years ago, half of humanity had never made a phone call and only 20 percent of humanity had regular access to communications. Today 70 percent of humanity can place a telephone call – or, more likely, send an SMS message – to the secretary general of the United Nations or most anyone else. For the first time in history, the majority of humanity is connected and has a voice.

Adults in Western culture fail to appreciate the momentous nature of this change, because our mindsets are tied to lumbering legacy technologies, like PCs and laptops. But in most countries, and for virtually all youth, the way to maintain your social network and run your business is by cell phone.

Digital connections allow public services to be transformed. In much of Africa, health workers survey the spread of disease, advise expectant mothers, and coordinate health services by digital messaging over cell phones. In tests, the digital system is both ten times faster than the old paper system – allowing health workers to nip epidemics in the bud – and ten times cheaper, despite the fact that phones cost more than paper.

Governance is also being transformed. Not only have the heads of governments been brought down by SMS-organized protests but multilateral organizations, such as the World Trade Organization, have been brought to account as well. More subtle, but perhaps even more important, the traceable nature of digital transactions means that banking and government services offered by cell phone are more transparent and accountable than the older systems. An example of this capability in action is that most captured terrorists have been identified through cell phone transactions.

Perhaps most important, connection means improved efficiency and greater wealth. In some parts of Africa and South Asia, banking is done by moving around the money in cell phone accounts. People pay for vegetables, or taxi rides, by SMS. Because remanufactured cell phones cost $10 in the developing world and incoming messages are free, every stratum of society is connected. Day laborers, for instance, no longer hang around on street corners waiting for jobs; instead, job offers arrive by SMS from a computerized clearing house. The International Telecommunications Union estimates that in the poorest countries each additional cell phone adds $3,000 to the GDP, primarily because of the increased efficiency of business processes.

The human race finally has a working nervous system, and the poor and disenfranchised are for the first time beginning to make themselves heard and felt. To accelerate this process, at MIT we have established the Program for Developmental Entrepreneurship (web.mit.edu/de), which helps form, fund, and scale in-country efforts that leverage these new capabilities. The possibilities opened up by humanity's new nervous system are unprecedented and reason for great optimism.

Emergent Democracy and Global Voices

JOICHI ITO

Founder and CEO of Neoteny.

I am optimistic that open networks will continue to grow and become available to more and more people. I am optimistic that computers will continue to become cheaper and more available. I am optimistic that the hardware and software will become more open, transparent, and free. I am optimistic that the ability to create, share, and mix works will provide a voice to the vast majority of people.

I believe that the Internet, open source, and a global culture of discourse and sharing will become pillars of democracy for the 21st century. Whereas those in power – as well as terrorists, who are not – have used broadcast technology and the mass media of the 20th century against the free world, I am optimistic that the Internet will enable the collective voice of the people, and that it will be a voice of reason and goodwill.

Humanity's Coming Enlightenment

LARRY SANGER

Cofounder of Wikipedia.

I am optimistic about humanity's coming enlightenment.

In particular, I am optimistic about humanity's prospects for starting exemplary new collaboratively developed knowledge resources. When we hit upon the correct models for collaborative knowledge collection on-line, there will be a jaw-dropping, unprecedented, paradigm-shifting explosion in the availability of high-quality free knowledge.

Over the last few years, I have received an increasing amount of mail from researchers who want to build dynamic, efficient, and enormous new knowledge resources that follow the wiki model. I believe that researchers are drawn to the wiki model because they naturally love several ideas suggested by the model: working closely with large numbers of their colleagues spread over the world; updating shared knowledge on the fly and avoiding costly duplication of labor; presenting knowledge systematically and in all its glorious complexity; and providing clear and compelling free access to important knowledge of their fields to a world that, in many cases, desperately needs such access. These features make the wiki model exciting.

Researchers – scholars, scientists, professionals of all sorts, and, indeed, all folks who love books – are drawn to the wiki model of strong collaboration in growing numbers. It's an accident of history that methods of strong collaboration on-line began among programmers and then spread to a completely egalitarian community – Wikipedia and its many imitators. The next step in the evolution, which we are now witnessing on many fronts at once, is the adoption of wikis and similarly dynamic and efficient 'Web 2.0' methods of strong collaboration, by knowledge workers, or (as in the case of my new project, the Citizendium) huge open communities that are gently guided by experts.

I think the most fantastic knowledge resources of the near future will be not encyclopedias or textbooks but brand-new sorts of resources – resources never before possible because they require the participation of thousands, or even millions, of users. What, for example, can that quantity of people do with millions of free books at their fingertips?

Assuming, as I do, that expert-guided but public participatory free knowledge projects are feasible, and that after becoming convinced of their tremendous value millions of intellectuals worldwide will spend significant amounts of time developing them together, my considered view is that we are approaching a kind of intellectual revolution:

- The time spent in library, reference, and literature research will be shortened by orders of magnitude, as increasingly detailed indexes and various kinds of maps of the literature are made available, as brilliant new search methods become available for the entire contents of enormous libraries, and as literature reviews (and similar resources) at all levels of specialization are constantly updated.
- Indeed, due to these coming sea changes in the way the results of research are accessed, we might well see new and more efficient methods of presenting novel research to the world than the traditional peer-reviewed research paper. (I

suggest nothing in particular here, but as a general point it seems not unlikely.)

- In many fields, especially in the humanities and social sciences, what today appears to be 'cutting edge' thinking will be placed into an easily accessible historical context. It will appear as so much scholarship properly does appear: old ideas, warmed over. I think the embarrassing ease of access to historical precursors and related work will lead scholars in these fields to focus on hard study, consolidation, and systematization – and maybe even teaching (actually, I don't think that; that would be too optimistic).

- There will be – as is already happening – a global research community coextensive with the Internet. This, even more than the advent of the Internet itself, has the potential to bring scholars from developing nations into the world of research as nothing has before. A well-educated, well-plugged-in intelligentsia from every uncensored place on the map could have many remarkable effects.

- Perhaps more important than any of the above, the ease with which information is accessed by teachers and students will require a complete and long overdue rethinking of the methods of education. What happens to education in a world in which not only some questionable information is available pretty quickly – as is now the case via Google and Wikipedia – but in which the most 'officially' reliable information is available, practically instantly, to everyone? What would teachers do with their students if classes were held every day in the middle of the largest library in the world?

Those are my expectations, and to have such expectations is, of course, to be very optimistic. I should add that it seems that the management of these information resources will have tremendous power, because of the tremendous value of their resources. So I hope that these managing bodies or persons will use their power according to the love of free speech, Western democratic republican principles of governance, and the rule of law.

Metcalfe's Law of Minds

CHRIS ANDERSON

Editor-in-chief of *Wired*; author of *The Long Tail: Why the Future of Business is Selling Less of More.*

Our species is unmatched in its ability to use communications to spread learning across populations, allowing us to get smarter and more capable far more quickly than evolution alone would allow. What makes me continually hopeful is that those tools of communications continue to get so much better so much faster. Anyone who can explore Wikipedia and not be both humbled and filled with confidence in the collective potential in the people all around us is a cynic indeed. And we've only scratched the surface of such networked intelligence.

Metcalfe's Law (formulated by Robert Metcalfe, coinventer of the Ethernet networking standard) says that value of a network grows with the square of the number of nodes. Today's Web, which is as much about contributing as about consuming – two-way links, as opposed to the old one-way networks of broadcast and traditional media – allows the same to apply to people. Connecting minds allows our collective intelligence to grow with each person who joins the global conversation. This information propagation process, which was once found in just a few cultures of shared knowledge, such as academic science, is now seen on-line, in everything from hobbies to history. The result, I think, will be the fastest increase in human knowledge in history.

This morning I was explaining Moore's Law and the magical power of this continuous learning curve to a nine-year-old. 'Will it ever end?' he asked. 'I don't see why it should,' I answered. That's optimism for you.

Altruism on the Web

DAN SPERBER

Social and cognitive scientist; *directeur de recherche*, CNRS, Paris; author of *Explaining Culture: A Naturalistic Approach*.

Had the question been 'What are you pessimistic about?' I would have answered: If there is any progress in human wisdom (and, yes, I suppose there is) it is pathetically slow, while ever faster technological advances provide the means for self-righteous, unwise people with power, wealth, or charisma to cause greater and greater havoc. I don't, alas, have any equally broad and compelling reasons to be optimistic about the future of humankind. Humans, however, are full of surprises, many of them excellent, arousing new hopes every day.

'From each according to his ability, to each according to his needs.' So did Marx define communism. Outside of narrow kinship or friendship groups, this kind of altruistic sharing of resources has hardly ever been encountered, and it is not difficult to understand why: Such a utopia, however attractive, is quite impractical. Yet with the advent of the new information technologies – and, in particular, of the Web – a limited form of informational 'communism' that no one had predicted has emerged and is fast developing. A vast array of technological, intellectual, and artistic creations, many of them of outstanding quality, are being made freely available to all according to their needs, by individuals working according to the best of their abilities and often seeking self-realization more than recognition. I have in mind the freeware, the wikis, the open-source programs, the open-access documents, the millions of blogs and personal pages, the on-line text, image, and music libraries, the free Web sites catering to all kind of needs and constituencies. Who had been optimistic enough to expect not just the existence of this movement but its expansion, its force, its ability to rival commercial products and major businesses and create new kinds of services – blogs, for instance – of great social and cultural import even if of limited economic value?

Cynics or realists – call them what you want – might say: Economic benefit is still the main force driving innovation. Gifted, disinterested amateurs – if that is truly what they are – are a welcome anomaly spurring competition, but what matters to the end user is the utility of the product. A cheaper product – and *a fortiori* a free one – is preferable, everything else being equal, but businesses, by providing extra quality worth the cost, make sure that everything is rarely equal. So let us praise innovation wherever it comes from, paying the price when justified and mouthing a word of praise when it comes free. But let us not read too much ('informational communism'? Give me a break!) into a probably ephemeral sociological oddity. As many others have noted, the economics of information are peculiar, if only because you can give away information without losing it, and you may gain from giving it away as much or more as from receiving it. Applying a standard economic model to the movement of information on the Web may not be the best science (actually, applying a standard economic model to standard economic situations may not be the best science either).

I am optimistic about the development of both individual and collective forms of altruism on the Web. Moreover, I believe that what we see on the Web has more diffuse counterparts in society at large. The Web is a network of networks, where, at every individual node, many communities overlap and where local allegiances have at best a weak hold. The World Wide Web is the most dynamic and visible manifestation, and a driving force, of a world that is itself becoming one wide web. In this world, more and more altruistic acts – acts that had in ancestral times been aimed just at one's kin and later extended to tribe, sect, or country – may now, out of a sensible sense of common destiny, be intended for the benefit of all. No Hallelujahs, however. If our destiny indeed is ever more common, it is because we all stand to suffer from the misdeeds of a few as much as to benefit from the generous actions of many.

The End of the Commoditization of Knowledge

ROGER C. SCHANK

Cognitive psychologist and computer scientist; founder of Engines for Education, Inc.; author of *Making Minds Less Well Educated Than Our Own*.

Fifteen years ago, I was asked to join the board of editors of *Encyclopaedia Britannica*. In short order, I learned that these editors saw themselves as guardians of knowledge. They knew what was true and what was important, and only knowledge that fit those criteria would be in their encyclopedia. I asked if the encyclopedia could be say, ten times bigger, economic issues aside, and they said no, the right information was already in there. I started to explain that the world as they knew it was going to change before their eyes and they would soon be toast, but they didn't understand.

I have had similar conversation with newspaper editors, librarians, heads of testing services, and faculty at top universities. Like the Britannica folks, they see themselves as knowing what is true and what is not and what is important and what is not. I am optimistic that this is soon all about to change.

What I mean by 'this' is the era we have lived in ever since the invention of the book (clearly including the time when knowledge was contained in scrolls). In this era, knowledge is a commodity, owned and guarded by the knowledge elite and doled out by them in various forms that they control, like books, newspapers, television, and schools. They control who can get access to the knowledge (through admission to elite schools, for example) and exactly what knowledge matters

(through SATs but also through intellectual publications that true knowledge owners would be embarrassed not to have read.)

We are beginning to see the change in all this now. If anyone can take Harvard's courses on-line, then one wonders why one has to go to Harvard. Elite universities have struggled with this new world, but eventually people will take whatever course they want from whomever they want, and a real competition about quality will take place.

We no longer have only three TV networks, so more points of view are available, but the cost of running a TV station is still high and there are barriers to entry and there is still the idea that a TV station broadcasts all day even if it has nothing to say. Soon this too will disappear. YouTube is just the beginning.

Today print media is being challenged by on-line material, but it is still prestigious to publish a book, and newspapers still exist. More important, schools still exist. But they are all going away soon. There is no need to buy knowledge when it is available for free, as newspapers are learning. When everyone has a blog and a Web site, the question will be whose information is reliable and how to find it. No one will pay a dime. Knowledge will cease to be a commodity.

I consider this to be a very good thing. I believe that those who own and dispense knowledge have turned off at least one and maybe more generations of thinkers, by telling them that all the important ideas are known and it is sad that you just don't know them. Religions have operated on this principle of knowing what is in the sacred scrolls for a very long time. Schools have acted similarly. Soon no one will be able to claim they know what is true, because people will be able to create debates for themselves. Google has helped make this happen already, but you ain't seen nothin' yet.

More important, the size of information will change. Today the size is a book or an article or a lecture or a course. But soon it will be a sound bite (or a paragraph). Yes – those terrible sound bites that the owners of knowledge like to complain about. Nuggets of knowledge will win because they always have won. People used to have conversations, sound bite followed by sound bite, directed toward a mutual goal held by the conversationalists. This will soon be possible on-line. You'll able to start a conversation with (an electronic version of) anyone who wants to offer this service. We'll be able to get information in the size that we got it thousands of years ago – a size we can process and respond to. No more blank-eyed listeners as people ramble on. Information will find us, and we will express our thoughts back. Knowledge will mostly be free and the owners of knowledge will need to go into another line of work. Knowledge will cease to be a controlled commodity.

Metacognition for Kids

GARY F. MARCUS

Director of the Infant Language Center, New York University; author of *The Birth of the Mind*.

We can use the discoveries of cognitive science to improve the quality of education in the United States and abroad. To do this, however, we need to radically rethink how our schools work. Since at least the time of the Industrial Revolution, the main emphasis has been on memorization, force-feeding our children with bite-sized morsels that are easily memorized – and quickly forgotten. (Recall the words of Charles Dickens' stern schoolmaster Mr. Gradgrind, 'Now, what I want is, Facts. Teach these boys and girls nothing but Facts . . . Plant nothing else, and root out everything else.') I am not sure it ever served a purpose for children to memorize the capitals of all fifty states (as I failed to do in junior high school), but in the age of Google, continued emphasis on memorization is surely largely a waste of time.

Five decades of cognitive science have taught us that humans are not particularly good memorizers – but also that we as a species have bigger fish to fry. Hamlet may have marveled that humans were 'noble in reason . . . infinite in faculty,' but experimental psychologists like Daniel Kahneman and the late Amos Tversky have shown that humans are actually often poor reasoners, easily fooled. The average person tends to have a shaky grasp of logic, to believe a lot of what he or she hears unreflectively, and to be overly confident in his or her own beliefs. We tend to be easily fooled by vivid examples and to notice data that support our theories – while forgetting about or ignoring data that refute them. Yet I cannot recall a single high school class on informal arguments, how to spot fallacies, or how to interpret statistics. It wasn't until college that anybody explained to me the relation between causation and correlation. In the age of the Internet, our problem is not that children can't find information but that they can't evaluate it.

What today's children need is not so much a large stock of readily Googleable information as a mental toolkit for parsing what they hear and read. As the old saying goes, it is better to teach a man how to fish than to simply give him fish; the current curriculum largely gives children fish, without teaching them a thing about how to fish for themselves.

How to teach children to fish for themselves? I would start with a course in what cognitive scientists call *metacognition*, knowing about knowing. Call it 'The Human Mind: A User's Guide,' aimed at, say, seventh-graders. Instead of emphasizing facts, I'd expose students to the architecture of the mind, what it does well, and what it doesn't. And most important, how to cope with its limitations, to consider evidence in a more balanced way, to be sensitive to biases in our reasoning, to make choices in ways that better suit our long-term goals. Nobody ever taught me about these things in middle school (or even high school), but there's no reason why they couldn't be taught. In time, I expect they will.

The Immeasurables

SHERRY TURKLE

Abby Rockefeller Mauzé Professor of the Social Studies of Science and Technology in the Program in Science, Technology, and Society, MIT; author of *Evocative Objects: Things We Think With*.

An eight-year-old sits braiding the hair on the tail of her My Little Pony doll, completely absorbed in the job. The shining plasticized hair is long and resilient; she plays with it for hours.

She starts by taking the tail and dividing it into three pieces, which she braids together. Then she undoes that braid and begins to nest layers of braids. She divides the tail into nine pieces and braids each group of three until she has three braids, and then takes these three braids and braids them together. After a while, the girl is starting with twenty-seven pieces, braiding them first into nine, then into three, then into one. The girl is playing with My Little Pony, but she is thinking about recursion.

The 'eight-year-old' is actually one of my MIT students telling a story of her childhood. For the past thirty years, I have begun my courses at MIT by asking my students to write about an object that was important to them on their path toward science. What they have had to say testifies to the importance of objects in the development of a love for science – a truth that is simple, intuitive, and easily overlooked. And it is cause for optimism, because it offers a hopeful note as we face our national crisis in science education.

As we argue about testing and standards, about the virtues of digital tools, about whether or not to move to on-line courseware, we have a tendency – as in any emergency – to look for salvation in the next new thing or things. In this case, these next new things are testing, measurement, and the computer itself as a way to provide educational solutions. But we can also look back to the last things that worked, and one of the things that has always worked in getting young people interested in science has been object passions. From my very first days at MIT, in 1976, I met students and colleagues who spoke about how as children they were drawn to science by the mesmerizing power of a crystal radio, by the physics of sand castles, by playing with marbles, by childhood explorations of air-conditioning units.

Certain trends are apparent as I look at the objects that have drawn children to science over the past thirty years. One is an interest in 'transparency.' Through the mid-1980s, MIT students wrote about being drawn to science by radios, vacuum cleaners, wooden blocks, and broken telephones – things to take apart and put back together again. By the end of the 1980s, the emphasis had shifted to objects that are investigated through the manipulation of program and code. Yet even with the passage from mechanical to electronic, and from analog to digital, students expressed a desire to get close to the inner workings of their machines. Even with machines that are increasingly opaque – with a printed circuit board, where one can no longer 'open the hood and look inside' – young people with a scientific bent continue to search for at least a metaphorical understanding of the mechanism behind the magic. And they find it.

Beyond seeking a way to make any object transparent, students extol the pleasure of materials, of texture – what one might call 'the resistance of the real.' For one such student, geology became real through her childhood experience of baking a chocolate meringue: 'Basic ingredients heated, separated, and cooled equals planet.' A thirteen-year-old looks up at the motion of his fly line while fishing with his father and is reminded of drawings of long, continuous, flowing lines he had made in algebra class: 'I realized that the motion of my hand had a very direct effect on the movement of the line, much in the same way that the input to a function produced a given output. Without any formal understanding of the physics involved, I was able to see the fly rod as representing a given function for which I was the input . . . From this point on, the fly rod was my metaphor for understanding function in mathematics.'

Young scientists are encouraged by a personal experience with an object they can understand and with which they can tinker. Playing with objects in their own way leads children to build a personal scientific style. There has been no simple migration to a new digital world. Children grow up in many worlds – they are seduced by the virtual but always brought back to the physical, to the analog, and, of course, to nature.

Science is fueled by passion, a passion that often attaches to the world of objects much as the artist attaches to his paints, the poet to his words. Young scientists are encouraged by a personal experience with an object they can understand and with which they can tinker. Playing with objects in their own way leads children to build a personal scientific style. Putting children in a rich object world is essential to giving science a chance. At a time when science education is in crisis, giving science its best chance means guiding children to objects they can love. Children will make intimate connections, connections they need to construct on their own.

One of the things that keeps educators and parents from valuing children's object passions is the fear that children might become trapped in objects and some might come to prefer the company of objects to that of other children. But even if the objects in the life of some young scientists do begin as objects that compel because they are more reassuring than people, such objects – from the periodic table of the elements (because it offers an image of perfect and reassuring organization) to LEGO blocks (because they offer a way to create perfect and reassuring symmetries) can become points of entry to larger, transformative experiences of understanding and confidence, very often at the point they are shared.

It seems wise to attend to young scientists' romance with

objects. If we do so, we are encouraged to make children comfortable with the idea that falling in love with things is part of what we expect of them. We are encouraged to introduce the periodic table as poetry and LEGOs as a form of art.

In *Mindstorms: Children, Computers, and Powerful Ideas*, Seymour Papert writes of falling in love with the gears of a toy car his father gave him when he was two. Fascination with those gears led to fascination with others. He played with gears in his mind, and mathematics began to come alive for him. He fell in love with the gears and he fell in love with science, all at the same time. Papert makes the point that if anyone had tried to measure what was happening to him as this inner explosion of affect and cognition was occurring, they would have found nothing to measure.

I am optimistic because a conversation about objects reminds us that just because we can't take a measurement doesn't mean that something important is not occurring. Too often when we can't test, we end the conversation. My students' voices make me hopeful because they serve as a reminder that the limit of testing is not the limit of inquiry. It can mark the moment when we turn directly to a child, when we put our deeper intelligence to work. We can learn what motivates and what inspires. If we insist on one-kind-fits-all curricular programming that takes children away from the idiosyncratic objects they are drawn to, we could miss a child who makes Cs and Ds in math and science but goes on to develop an abiding love for designing complex systems because he has connected with LEGOs and a personal computer. We could miss a child who doesn't think of herself as a science student even as she silently absorbs everything she can learn from chemistry experiments that create purple smoke. We might not count as learning the lessons that come with braiding a pony's tail, casting a fly rod, or baking a meringue.

I am reminded daily of these object passions in the students I teach; I am optimistic as I begin to sense the political and philanthropic will that could enable these passions to find their voice in education.

The Coming Revolution in Science Education

LEON LEDERMAN

Physicist and Nobel laureate; director emeritus of Fermilab; coauthor (with Dick Teresi) of *The God Particle: If the Universe is the Answer, What Is the Question?*

I am optimistic about science education. I know – I should have a psychological check-up, be tested for delusional fantasies, have my PhD revoked in a public ceremony with the breaking of my pencils. After all, in 1983 we were officially declared 'A Nation at Risk.' Commissions galore sported such titles (the 1999 Glenn Commission, 'Before It Is Too Late;' the Education Commission of the States, 'No Time to Waste'). The Hart-Rudman Commission came close to recommending that the budgets of the Defense Department and HEW be swapped. Eminent CEOs like Bill Gates (Microsoft), Craig Barrett (Intel), Louis Gerstner (IBM), Norman Augustine (Lockheed Martin), all corporations that depend on rationality for their profits, agree that our system of independent states, 15,000 school districts, 26,000 high schools, and so on has failed cata-strophically to educate our students for life and work in the 21st century.

But the good news is that the import of our failed educa-tional system as it affects our health care, our economy, our culture, and our status in the globalized world, is finally becom-ing clear – clear to parents, clear to economists, clear (*gasp!*) even to members of the Congress.

Somehow we have created a *Sputnik*-like climate, warning of a powerful enemy. Not the Soviets but an even more worthy opponent in a war we must declare and win: the War on Ignorance. Out of *Sputnik* came the National Defense Education Act of 1958, NASA, and a renewed determination to modernize science education, but also the conjugate commu-nication skills, including in foreign languages, and a need for public science literacy.

Can one imagine parents who know that the earth revolves around the sun? And, for all of the *Edge* scientists, an audience who might actually read their books? Now, there is wild opti-mism!

High-Resolution Images of Earth Will Thwart Global Villainy

CHRIS DiBONA

Open Source Programs manager at Google, Inc.; editor of *Open Sources: Voices from the Open Source Software Revolution* and *Open Sources 2.0*.

I am not so much of a fool as to think that people will stop the careless destruction of flora and fauna for personal, corporate, national, or international gain. I do believe that the advent of rapidly updating, citizen-available, high-resolution imagery will remove the protection of the veil of ignorance and secrecy from the powerful and exploitative among us.

No one can tell you that clear-cutting a forest isn't so bad, if you can see past the half-acre of preserved trees into the desert-like atmosphere of the former rain forest. No one can tell you that they are not destroying villages in Sudan, if you can view the burned-out homes of the slaughtered. No one can tell you that the impact of a dam is minimal, as humanity watches countless villages being submerged. No one can paint a war as a simple police action, when the results of the carpet bombing will be available in near real time on the Internet.

We have already started down this path, with journalists, bloggers, and photographers taking pictures and uploading them to Web sites for people to see. Secrecy of this kind is dying, but it needs one last nudge to push our national and international leadership into a realm of truth unheard of to date.

With sufficient resolution, many things will be clear to all of us: troop movements, power-plant placement, ill-conceived dumping, or just your neighbor building a pool. I am optimistic enough to think that the long-term reaction to this kind of knowledge will be proper management and monitored phase-out of the unwanted. I am not as optimistic about the short term, with those in power opting to suppress this kind of information access, or worse, acting on the new knowledge by bringing to a boil conflicts that have simmered for years.

Can our leaders stand before us and say something is not occurring, if we can see, via our low-earth-orbiting eyes, that it is? Only the truly deluded will be unable to see, and then perhaps we can remove them from power. A more honest existence, with humankind understanding the full global impact of its decisions, is in our future, if we can reach it. It is likely to be a rough ride.

Transparency Is Inevitable

DANIEL GOLEMAN

Psychologist; author of *Social Intelligence*.

I live in a bowl-shaped valley on the edge of the Berkshire hills in New England. The prevailing winds come from the southwest. As it happens, a coal-burning electric plant sits in the dip next to the Holyoke Range at the southern edge of the valley, perfectly placed to fill the air with its unsavory mix of particulates. The plant is a dinosaur that, because of various regulatory loopholes, has been able to dodge costly upgrades that would make its emissions less toxic.

Nobody seems to mind. True, the head of pulmonary medicine at the local medical center bemoans the toll of the Mt. Tom power plant's particulates on the respiratory tracts of those who live in the valley, particularly the children. But the plant's operators blithely buy carbon-pollution credits that let it avoid the expense of upgrading its scrubbers.

The indifference of those of us whose respiratory systems routinely become inflamed, I'm convinced, is largely due to a failure of collective awareness. As we join the throngs in the waiting room of the local asthma specialist, we make no connection between our being there and that smokestack, nor between our use of electricity and the rate at which the smokestack belches its toxins.

I'm optimistic that one day the people in my valley will make those connections. Inexorably, the toll that our collective habits of consumption wreak on our environment and our health will become obvious.

My optimism does not hinge on the promise of some new technological fix or scientific breakthrough; rather, it stems from the convergence of market forces with off-the-shelf possibilities from a field that has already reshaped our lives: information science.

'Ultimately, everybody will find out everything,' as they say at the Googleplex, Google's corporate headquarters harboring perhaps the world's densest aggregate of specialists in data mining and other applications of information science. Information science, the systematic organization and meta-knowing of all we know, has been steadily increasing the sheer quantity of what each of us can find out.

One branch of this science, medical informatics, has transformed medicine by making instantly available to physicians a vast array of background data on their patients' condition, history, prognosis, and best treatment. One of the more hopeful applications of information science would be what we may call consumer informatics, which would do for the marketplace something akin to what's being done for medicine – make visible the elusive links between what we buy and do and the impact on our bodies and on the nature of the processes that support these activities.

Take, for example, the hidden costs of a T-shirt. John C. Ryan and Alan Thein Durning's *Stuff: The Secret Lives of Everyday Things* (1997) deconstructs ordinary products into the chemical consequences of their manufacture. Chemical by-products of textile dyes include chlorine, chromium, and formaldehyde; because cotton resists coloring, about a third of the dyes fail to adhere and so end up in wastewater. There are correlations between high levels of dye run-off in groundwater and rates of leukemia in local children.

For that reason, Bennett & Company, a supplier of clothes to companies like Victoria's Secret and Polo.com, has formed a partnership with the dyeworks that supplies its plants in China. The partnership allows the clothes manufacturer to ensure that the wastewater from the dyes it uses will go through a series of cleansing pools before returning to the water supply, rather than simply being dumped.

Here's the challenge for information science: Quantify the environmental and health impacts of the standard processes used in manufacturing goods. Then tag a given product on a store shelf, so that consumers can weigh its benefits against its harmfulness. Let us know which T-shirt has what consequences.

Large retailers use an electronic tagging system for inventory control, which lets a store manager know the history of every item on the shelves, including the factory where it was made. One next step would be to document the manufacturing practices at that factory, and so tag the item with its environmental/public-health legacy.

Here market forces may assist, encouraging companies to provide such information in the interests of competitive advantage. Some marketers have long touted product virtues in marketing. For example, Cascade toilet paper claims manufacturing methods that use no chlorine and 80 percent less water than the industry average. Some energy providers offer an option to buy electricity from renewable sources, like wind power. If companies themselves do not take such steps, there are alternatives. Today anyone can go into a store and, using a Palm Pilot to scan the bar code, be whisked to a Web site that could reveal information about that product's level of virtue – say, in terms of toxic chemicals unleashed during its manufacture.

But for such a Web site to have both credibility and teeth, there needs to be a sustained collaboration between information-science engineers, chemists, physicists, environmental scientists, and public-health and other medical specialists, as well as manufacturers. The mass of data would be potentially immense; information science sorts out the signal from the noise or reorganizes noise into signal.

The task is daunting, but I'm optimistic that we're heading toward the day when we will be able to vote with our wallets every time we go shopping.

Power Is Moving to the Masses – As a Market

ESTHER DYSON

Editor, *Release 1.0*; trustee, Long Now Foundation; author of *Release 2.0.*

Many of the venture capitalists I know are turning to environmental and energy investments; the more adventurous are looking at health care (not just drugs), low-end PCs, products for the masses. They are funding training schools in India – for-profit – rather than just donating to legacy universities in the United States. The watchword is 'sustainability.' In plain English, that means making a profit so that more profit-seeking investors will enter and enlarge the market.

They have a variety of motives, ranging from altruism and the desire to solve problems to a need for recognition (or belief in the profitability of doing so), but whatever the motive, the result is cause for optimism. Millions of investors and entrepreneurs will apply their resources and talents to improving products, distribution systems, training and education and health-care facilities targeted at the billions of people at the bottom of the pyramid. In our fluid world of competition and fast-spreading information, some people will still get rich by being first and smartest, but most will get rich by implementing well and serving broader markets. For the first time in history, power is really moving to the masses – not as a power bloc but as a market.

Of course, all those billions of people will also be producers, and a broader spread of education and productivity tools, from water pumps to cell phones and PCs, will enable them to join the world economy as productive people.

Yes, this is an optimistic view and it won't all be simple, but the forces – from human dignity to human greed – are aligned. You asked for the optimistic view, and optimism will help make it happen!

Capitalism Is Aligning with the Good

JASON McCABE CALACANIS

Founder of the *Silicon Alley Reporter*; chairman of The Weblogs, Inc. Network.

Capitalism is becoming more closely aligned with the forces of good (e.g., philanthropy) than with the forces of greed. As the polarization of wealth peaked, the press and public became obsessed with the 'Greed is good' meme in the eighties and the 'Rules don't apply to the Rich' nineties (think Enron and Worldcom). However, the real story was brewing, and we read it first on *Edge* and witnessed it in Ted Turner's gift to the United Nations.

The most successful businesspeople in the world have decided to dedicate their brains and bank accounts to fixing the world, leaving politics and politicians on the sidelines. Bill Gates, Warren Buffett, Richard Branson, John Doerr, and Pierre Omidyar – among many others – are demonstrating that the true goal of winning is giving. The brass ring has moved from private aviation and mega-yachts to making a mega-pledge at Bill Clinton's annual summit.

Edge's ongoing discussions on happiness are clearly contributing to the trend. As psychologist Martin E. P. Seligman noted in his 2004 *Edge* feature 'Eudaemonia, The Good Life': 'The third form of happiness, which is meaning, is again knowing what your highest strengths are and deploying those in the service of something you believe is larger than you are. There's no shortcut to that. That's what life is about.'

Individuals are Empowered in a Knowledge-Driven Economy

JUAN ENRIQUEZ

CEO of Biotechonomy; founding director of Harvard Business School's Life Sciences Project; author of *The Untied States of America.*

Our freedom to create, to work, to fundamentally alter our circumstances, is unprecedented. For better and worse, science and technology provide ever greater power. Individuals and small groups can leverage this power to set their own rules, make their own lives, establish their own boundaries. Paradoxically, this is leading to massive global networks and ever smaller countries.

Don't like your country, or your neighbors? Had enough of the religious, nationalist, or ethnic fanatics nearby? Groups that like to speak obscure languages, that want to change the heroes in the grammar-school textbooks, that advocate a different set of morals, are increasingly able to do so. Borders and boundaries in rich and poor countries alike, are ever shifting. There are demands for autonomy or outright separation in the United Kingdom, France, Belgium, Netherlands, Austria, Germany, Italy, Spain.

It used to be the case that the bright had to leave India, Pakistan, China, or Mexico to make a living or have a global impact. No more. There are within such countries ever more powerful enclaves and zip codes where there is a concentration of smarts and entrepreneurship. If their neighbors let them, these groups and networks flourish. If not, they separate or they leave.

As long as you can become and remain a part of the network, the power of place matters ever less. Who your parents were, where you were born, is irrelevant, as long as you have access to and interest in education, technology, science, and networks. Every time you open a science magazine, research a new material or gene, map a brain, ocean, or piece of the universe, there is a smorgasbord of opportunities to learn and build something new, to create, to accumulate enough prestige, wealth, or power to fundamentally change many lives.

A knowledge-driven economy allows individuals to lead millions out of poverty in a single generation. Many within the biggest nations (China and India) as well as the smallest (Singapore and Luxembourg) can thrive. You no longer need to take what your neighbor has in order to survive. You can thrive by building something complementary, by thinking up something better. Knowledge – unlike land, oil wells, gold mines – is ever expanding. Your success depends less and less on taking what the other has. You can build or make your own.

Sometimes the real and the cyber merge and cross over. Second-lifers, inhabitants of a virtual world, are free to become whatever they wish, wherever they wish. Their creations and wealth are increasingly crossing boundaries into the real world of corporate planning, art, and dollars. This is just a foreshadowing of real-world governance going forward, as the digital becomes the dominant language and means of generating wealth on the planet.

Throughout the world, your grandpa's favorite sports team probably moved, his flag changed, his job merged, moved, or morphed. All of this implies an accelerating set of shifts in allegiance and identity. Politicians and citizens who wish to preserve and protect their current country are well advised to pay attention to these trends, as more and more people have a choice and debate whether to become a more compact few. For the illegitimate or the slow, it will be harder to maintain boundaries and borders. There is little margin for error; each government and temporarily dominant party can screw up the whole. And the whole can be split very fast. But you have many options. You can fight to preserve what you love, or you can choose to build or inhabit an alternative space. Your choice.

Humans Will Learn to Learn from Diversity

DANIEL L. EVERETT

Researcher of Pirahã culture; chair of Languages, Literatures, and Cultures, and professor of linguistics and anthropology, Illinois State University.

I am optimistic that humans will finally come to understand that they can learn from other humans who are not like them. The supposed curse of the Tower of Babel – the diversity of languages and cultures – is perhaps our greatest hope for continued healthy occupancy of this rock we all share in our unforgiving universe. Sure, there are dangers in this diversity: It has sometimes led to murder and suffering. Diversity can all too easily be interpreted as incomprehensibility, inferiority, wrong-headedness. But I am optimistic that our species has grown tired of this view, and I am optimistic that groups we have heard very little from will motivate us all to learn new solutions to old problems in the coming years. However we define the group to which we belong – ethnically, geographically, linguistically, or nationally – I believe that 2007 could be the year in which we come to embrace a symmetry of status between groups and a cross-pollination of ways of living and ways of thinking about the world.

Let me say what I think it means for groups to learn from one another and why I am optimistic about it.

The world presents us all with similar problems at the level of biological and emotional need. We need shelter, food, companionship, affection, sex, and opportunities to develop our abilities, among other things. As humans, we also have intellectual and social needs that go beyond other species. We need affirmation, we need respect, we need to feel good about our lives, we need to feel as though we are useful, we need to feel optimistic. We need to know how to extract more meaning out of the world around us. And, especially, we need to learn to love more and tolerate more. But how do we learn these things? Where can we go for new solutions to the problems that still beset us? Anthropological linguistics can offer some suggestions. We can learn from the stories and values of smaller, overlooked groups, endangered peoples, and even those extinct peoples for whom we have records, about how to live more harmoniously in the world.

For example, when we look back to the extinct cultures of the Narragansett Indians, and to the early British settlers of northeastern America, we learn about their tolerance of difference. When the Puritan leaders of Massachusetts expelled Roger Williams from the colony in 1635 during a ferocious winter, they undoubtedly expected Williams to do the right thing and freeze to death. Williams had expressed views of tolerance and respect for others and against tenets of the church – views that offended them. But Williams was taken in by the Narragansett and spent the winter safely with them, learning about their language and their philosophy of tolerance, of which he was living proof. Williams' writings about these people later influenced the thought of Thomas Jefferson, and eventually the Narragansett philosophy seems to have influenced, though indirectly, the writings and thought of William James as he helped to develop American pragmatism, perhaps the only uniquely American contribution to world philosophy. It is a philosophy that evaluates ideas by their usefulness, and it is marked by tolerance of diverse ideas and rejection of the notion that any one group holds a monopoly on truth.

We've spent most of our time on this planet trying to homogenize it – to remove uncomfortable differences. But we are growing weary of this. I believe that this year the pain our species is inflicting on itself will surpass (for many of us, at least) what we're willing to bear. We will look for other answers. And we'll need to turn to humans who have mastered the art of contentment and peace and tolerance. These people are found in various parts of the world. Zen Buddhists are one example, but there are others.

My thirty years of work with the Pirahā (Pee-da-HAN) of the Amazon rain forest, for example, has taught me a great deal about their remarkable lack of concern over the future or the past and their pleasure in living one day at a time, without fear of an afterlife, with full tolerance for others' beliefs. The Pirahā know that people die, that they suffer, that life is not easy, through their daily struggle to provide food for their families without being bitten by snakes or eaten by jaguars, and from burying their young, victims of malaria and other diseases. But it doesn't dampen their joy in life, their love for one another, or their ability to look at death without fear and without need for the idea of Heaven to get them through this life – to them, the only life.

Religious truth lacks tolerance; it is a truth that wants to proselytize the world in order to eliminate diversity of belief. Western history has shown what that leads to. But peoples like the Narragansett, the Pirahā, the Zen Buddhists, and many others offer alternatives to homogenizing and destroying the diversity of the world. They show us how different people can solve the same problems of life in ways that avoid the by-products of the violent homogenization that has marked Western history. The Internet and the rapid dissemination of research in popular and professional forums, coupled with widespread disgust at some of the horrors our traditional cultural values have produced, can help us learn from them.

What is there to learn? Let me offer some examples from my own research among Amazonian peoples.

(1) *Cooperation.* I once thought it might be fun to teach the Pirahã about Western games, so I organized a field day, with a tug-of-war, a foot race, and a sack race, among other things. In the foot race, one fellow got out in front of everyone else. He then stopped and waited for the others to catch up so they could cross the finish line together. The idea of winning was not only novel but unappealing. We cross the line together or I don't cross it. The same went for the sack race. The tug-of-war was a joke – just guys keeping the slack out of the rope and talking. They loved it all, laughing and talking all day long and telling me what a good time they had. They taught me more than I taught them: You can have a great time and have everyone win. That's not a bad lesson – that's a fine lesson.

(2) *Pluralism.* The Pirahã, like the Narragansett and other American Indians, believe that you use your knowledge to serve both yourself and others in your community. There is no overarching concept of a truth to which all members of the society must subscribe.

(3) *Communalism.* The Pirahã seem to accept only knowledge that helps, not knowledge that coerces. Think of our English expression 'Knowledge is power.' The concept as practiced in most industrial societies is that 'Knowledge is power for me, as long as I have it and you don't.' But to people like the Pirahã, knowledge is to be shared. It is

power to the people, not power to a person. The Pirahã don't allow top-secret conversations. Every member of their society knows what the other members are doing and how they're doing it. There is a communal mind. There is freedom and security in group knowledge.

(4) *Toleration.* In Western society, we associate tolerance with education – the more you learn, the more you tolerate. But there is little evidence for this thesis in practice, and among some hunter-gatherer societies toleration of physical, mental, and religious diversity is much greater than in our so-called pluralistic Western societies. We need not all look alike, act alike, behave alike, or believe alike; in fact, we don't even have to pretend to do so.

In the 1960s, there was a similar optimism among my fellow hippies, as many of my generation went into fields like anthropology, literature, and science to learn more about diverse facts and truths and discover coping lessons for our lives. We are ready now for a new sixties-like exploration of diversity, and I am optimistic that we will do this. I am also optimistic that we will learn the simple and useful truths of cooperation, pluralism, communalism, and toleration and that no one Idea or Truth will bind us all.

Early Detection of Learning Disabilities or Difficulties

HOWARD GARDNER

John H. and Elisabeth A. Hobbs Professor of Cognition and Education,
Harvard Graduate School of Education; adjunct professor of psychology,
Harvard University; author of *Five Minds for the Future*.

When, at John Brockman's urging, I don the hat of scientific optimism, I think of the early detection of learning disabilities or difficulties, coupled with interventions that can ameliorate or even dissipate those difficulties. In the near future, we will be able to use neural imaging techniques to determine which infants or toddlers are at risk of having problems reading, writing, calculating, spelling, mastering spatial relations, mastering social relations, and the like. (We may even have genetic markers for those risk factors.) And I believe that the more specific the detection of the disorder (that is, which kind of reading problem, what sort of social deficit), the more easily we can devise interventions that address and perhaps eliminate the problem.

But as soon as I entertain this optimistic view, another, less happy scenario floods my consciousness. The same means of early detection can easily be put to malevolent purposes. First of all, we'll be identifying not just deficits that can be addressed but also ones that cannot. ('You'll get Alzheimer's disease at thirty-nine, and, alas, there is nothing we can do about it.') Second, we run the risk of stigmatizing children from birth. ('Oh, you're destined to be illiterate,' or 'You'll never be likable, because of your social deficits.') Moreover, we will likely soon turn not just to addressing deficits but also to trying to produce the ideal child – to enhance perfectly adequate abilities through genetic, neural, or pharmacological intervention. Not only does this intrusion go against human nature and fate as we have always known them, but it will also further privilege those who are already privileged and have disposable resources.

Thus, challenging the spirit of the *Edge* annual question, I maintain that to consider science apart from its use and its users – or its misuse and its abusers – is naïve. Accordingly I have to add some political remarks:

In recent years in the United States, we have seen many examples of how science can be distorted for political purposes. While many of us have protested this cynical attitude toward science, we have been ineffective in arousing the public or changing the behavior of the administration. And so, many citizens – particularly the young – have been tempted to turn their energies elsewhere. This ostrich-style behavior is risky. I recall a comment made to me several years ago by John Gardner (no relation), former secretary of health, education, and welfare and founder of Common Cause. 'There have never been so many young people in America involved in public service, community service, social entrepreneurship, and other efforts to promote the common good,' Gardner said. But he added somberly, 'This commitment won't add up unless these young people become involved in the political process. Because while they may be helping dozens or even hundreds of individuals, laws are being passed that harm thousands or millions of persons.' With reference to the cynical use of science, it is not enough to fret; we need political muscle if we are to see science reported objectively and not yoked to suit political purposes.

After the election of November 7, 2006, I feel a shade more optimistic about America. Increasingly, idealistic young people are engaged in politics and running, or considering a run, for office. America will be in better shape when the leaders and graduates of organizations like Teach for America or City Year meld their sense of public service with political involvement. It is vital that these engaged young persons understand what science is, and what it is not; it is essential that they not bend science to suit their ideological purposes. This is not to say that these future leaders should abandon their values; rather, they should fully consider the weight of scientific evidence relevant to the social goals they seek to achieve.

The Human Response to Vast Change Will Involve Strange Bounces

JOEL GARREAU

Cultural revolution correspondent, *Washington Post*; author of *Radical Evolution: The Promise and Peril of Enhancing Our Minds, Our Bodies — and What It Means to Be Human.*

I am an optimist because I have a hunch that Mark Twain was right when he portrayed Huckleberry Finn as an archetype of human nature. In the pivotal moment of his novel, Huckleberry Finn considers struggling no longer against the great challenges arrayed against him. He thinks about how society would shame him if it 'would get all around that Huck Finn helped a nigger to get his freedom.' ('That's just the way: A person does a low-down thing, and then he don't want to take no consequences of it. Thinks as long as he can hide, it ain't no disgrace. That was my fix exactly.')

Huck decides right then and there to abandon a life of sin, avoid eternal damnation, and for once do the right thing by society's lights. He decides to squeal – to write a letter to Jim's owner telling her how to recapture her slave. Then he gets to thinking about human nature:

> I felt good and all washed clean of sin for the first time I had ever felt so in my life, and I knowed I could pray now. But I didn't do it straight off, but laid the paper down and set there thinking – thinking how good it was all this happened so, and how near I come to being lost and going to hell. And went on thinking. And got to thinking over our trip down the river; and I see Jim before me, all the time: in the day, and in the night-time, sometimes moonlight, sometimes storms, and we a-floating along, talking, and singing, and laughing. But somehow I couldn't seem to strike no places to harden me against him, but only the other kind. I'd see him standing my watch on top of his'n, 'stead of calling me, so I could go on sleeping; and see him how glad he was when I come back out of the fog; and when I come to him again in the swamp, up there where the feud was; and such-like times; and would always call me honey, and pet me, and do everything he could think of for me, and how good he always was; and at last I struck the time I saved him by telling the men we had smallpox aboard, and he was so grateful, and said I was the best friend old Jim ever had in the world, and the *only* one

he's got now; and then I happened to look around, and see that paper.

It was a close place. I took it up, and held it in my hand. I was a-trembling, because I'd got to decide, forever, betwixt two things, and I knowed it. I studied a minute, sort of holding my breath, and then says to myself:

'All right, then, I'll *go* to hell' – and tore it up.

It was awful thoughts, and awful words, but they was said. And I let them stay said; and never thought no more about reforming. I shoved the whole thing out of my head; and said I would take up wickedness again, which was in my line, being brung up to it, and the other warn't. And for a starter, I would go to work and steal Jim out of slavery again; and if I could think up anything worse, I would do that, too; because as long as I was in, and in for good, I might as well go the whole hog.

This line of rampant and pugnacious human perversity surfaces in our best stories again and again. In *Casablanca*, Rick is ensconced in a cozy world of thieves, swindlers, gamblers, drunks, parasites, refugees, soldiers of fortune, genially corrupt French police, and terrifying Nazis. Rick's cynicism is his pride; he sticks his neck out for nobody. His only interest is in seeing his Café Américain flourish. And then, of all the gin joints in all the towns in all the world, Ilsa walks into his. The rest of the film has him betraying his own cauterized heart in service of a higher purpose. As Rick says, 'It's still a story without an ending.'

The most phenomenally successful film series of the recent era – the *Star Wars, Harry Potter, Matrix,* and *Lord of the Rings* movies – are all driven by a faith in human cussedness, from

Han Solo's grudging heroism to little people with furry feet vanquishing the combined forces of Darkness.

If the ageless way that humans process information is by telling stories, what does our belief in this recurring story say about us?

It is an instinct that the human response to vast change will involve strange bounces.

This assessment of our species displays a faith that even in the face of unprecedented threats, the ragged human convoy of divergent perceptions, piqued honor, posturing, insecurity and humor will wend its way to glory.

The Future of Software

DAVID GELERNTER

Computer scientist, Yale University; chief scientist of Mirror Worlds Technologies; author of *Americanism: The Fourth Great Western Religion.*

I am optimistic about the future of software, because more and more people are coming out of the closet every month – admitting in public that they hate their computers.

Within the last month, I've heard three people shouting (or muttering) curses at their machines. One was a bona-fide software virtuoso! These particular three were ticked off about (1) an airline Web site that was so badly designed it was useless, (2) a commercial Web-site-building tool (bought for real money) that made it nearly impossible to build simple structures, and (3) a home PC that, despite reasonably sophisticated software countermeasures, was so junked-up with viruses that starting a word processor took five minutes.

The file systems and desktop and spreadsheets, the word processors and mailers and database programs we rely on are vintage 1984 or older. They're as obsolete as a 1984 PC. When I first described the 'empty computer' model in the early nineties, people thought I was crazy. Many still do – but fewer each year (and I guess that's progress). There was a larger jump in admitted cases of computer and software hatred in 2006 than in any previous year I remember.

Technologists who blandly assume that hardware will (somehow) keep getting better while software stays frozen in time are looking wronger every month. In the empty-computer world of the near future, your information assets have all been bundled up, encrypted, and launched into geosynchronous orbit in the cybersphere; computers are interchangeable devices for tuning in information. (If computers are so cheap, why does everyone need to carry one around? We don't make you carry a desk and chairs around with you; we can afford to provide chairs and flat surfaces wherever you need them.)

In the empty-computer world, it will take five minutes to upgrade to a new machine (throw the old one out, plug the new one in – your information stays in orbit, where it's always been); comfortable large-screen public computers will be available all over the place. And instead of expanding into a higher-and-higher-entropy mess, the Web will implode into a 'blue hole' – a single high-energy information beam that holds all the world's digital assets.

Gelernter's Law: The computer industry revolutionizes itself at least once a decade. We're nearly due for the next revolution.

Getting It All Wrong

STEVE GRAND

Artificial-life researcher; creator of Lucy, a robot baby orangutan; author of Creation: Life and How to Make It.

What I'm most optimistic about is the strong possibility that we've got everything horribly wrong. All of it. Badly.

Once, when I was a young child, I accompanied my father on a car journey around some back lanes in England. Dad wasn't familiar with the area, so I helpfully took the map from him and navigated. 'At the next crossroads, turn right. Then we'll come to a lane off to our left and a sharp bend. After the bend, turn right again.'

We turned right, passed the lane, followed the sharp bend and turned right. Easy. 'In a minute, we'll go past a pub on our right, then a pond, and then we take the second left.'

There was no pub, but there was a pond, so Dad turned as instructed. The next waypoint wasn't quite right, either, but the following two confirmed our route and I expertly navigated us from turn to turn. It was only after about half an hour of this that my father got a bit nervous about the number of anticipated bridges, churches, and rivers that had unaccountably disappeared since the map was published. As we sat staring helplessly at a field gate that should have been a major junction, it became obvious that something had gone badly wrong: When I'd taken over the navigating, we were actually on a completely different page of the map than I thought, and it was pure coincidence that enough of the landmarks matched the wrong page to convince us that we were still on track.

I learned a lesson from this. Science sometimes learns these lessons, too. Thomas Kuhn put it much better when he introduced the concept of a paradigm shift. Sometimes we manage to convince ourselves that we have a handle on what's going on, when in fact we're just turning a blind eye to contradictory information. We discard that information or ignore it (or can't get funded to look at it), because we don't understand it. It seems to make no sense, and a while goes by before we realize that the problem doesn't lie with the facts but with our assumptions.

Paradigm shifts are truly wonderful things. Suddenly the mists clear, the sun comes out, and we exclaim a collective 'Aha!' as everything begins to make sense. Why I'm so optimistic about science right now is that there are plenty of these 'Aha!' moments waiting in the wings, ready to burst onto the stage. We've got so much completely wrong, but we're starting to look at the world in radically new ways – dynamical, nonlinear, self-organizing ways – and a lot of our standing ideas and assumptions about the world are about to turn inside out, just as our much older religious ideas did during the Enlightenment.

My guesses for prime candidates would include quantum theory and our understanding of matter, but those aren't my field and it's not my place to judge them. My field is artificial intelligence, but I'm sad to say that this subject started on the wrong page of the map many years ago and most of us haven't woken up to it yet. We keep our eyes firmly on the route and try not to look to left or right, for fear of what we might see. In a way, Alan Turing was responsible for the error, because his first big idea in AI (which was that something vaguely reminiscent of human thought could be automated) turned out to be such a strikingly good one (for other reasons entirely) that it eclipsed his later, more promising ideas about connectionist systems and self-organization. Since then, the digital computer has dominated the AI paradigm, through failure after dismal failure.

My great white hope for AI lies in neuroscience. The only working intelligent machine we know of is the brain, and it seems to me that almost everything we think we understand about the brain is wrong. We know an enormous amount about it now, and just about none of it makes the slightest bit of sense. That's a good sign. It shows us that we've been looking at the wrong page of the map.

Let me illustrate this with a thought experiment: Suppose I give you a very complex system to study – not a brain but something equally perplexing. You discover quite quickly that one part of the system is composed of an array of elements of three types. These elements emit signals that vary rapidly in intensity, so you name them the alpha, beta, and gamma elements and set out eagerly to study them. Placing a sensor onto single examples of each type, you find that their signal patterns are distressingly random and unpredictable, but with effort you discover that there are statistical regularities in their behavior: beta and gamma elements are slightly more active than alpha elements; when beta elements are active, gammas in the same region tend to be suppressed; if one element of a given type changes in activity, its neighbors tend to change soon after; gamma elements at the top of the array are more active than those at the bottom – and so on. Eventually you amass an awful lot of information about these elements, but still none of it makes sense. You're baffled.

So allow me to reveal that the system you've been studying is a television set and the alpha, beta, and gamma elements are the red, green, and blue phosphor dots on the screen. Does the evidence start to fit together now? Skies are blue and tend to be at the top, while fields are green and tend to be at the bottom; objects tend to move coherently across the picture. If you know what the entire TV image represents at any one moment, you'll be able to make valid predictions about which elements are likely to light up next. By looking at the entire array of dots at once, in the context of a good system-level theory of what's actually happening, all those seemingly random signals suddenly make sense. 'Aha!'

The single-electrode recordings of the equivalent elements in the brain have largely been replaced by system-level recordings made by fMRI, but we still don't yet know what any of it means, because we have the wrong model in our heads. We need an 'Aha!' moment akin to learning that the phosphor dots belong to a TV set upon which images of natural scenes are being projected. Once we know what the fundamental operating principles are, everything will start to make sense very quickly. Painstaking deduction won't reveal this to us; I think it will be the result of a lucky hunch. But the circumstances are in place for that inductive leap to happen soon, and I find that tremendously exciting.

Isaac Newton once claimed that he'd done no more than stand on the shoulders of giants. He was being far too modest. It might be more accurate to say that he stayed down at child height, running between the giants' legs and exploring things in his own sweet way. It's that combination of artful playfulness and pan-disciplinary sources of analogy and inspiration that will turn our understanding of the world inside out. I'm optimistic about that.

Unraveling Beliefs

MAHZARIN R. BANAJI

Richard Clarke Cabot Professor of Social Ethics in the Department of
Psychology and Carol K. Pforzheimer Professor at the Radcliffe Institute
for Advanced Study, Harvard University.

If I were an average white woman living in the United States in
1850, I would already have been dead for ten years. Not an
ideal position from which to contemplate optimism about the
future, you say. But consider this: In the course of 150 years, the
life expectancy of this group catapulted from a dismal forty to
a respectable eighty years.

How can life expectancy, seemingly determined by biology
and the conditions of one's life, double so fast? Advances in science and technology are drivers of the rapid changes in
nutrition, medical care, and standards of living that account for
this doubling. But such advances were possible because of something else: changes in the mental states we call beliefs – beliefs
about the worth of a life, beliefs about what it means to be
happy, beliefs about health and prosperity, including about who
deserves it and who does not.

Many others in this volume will speak about specific accomplishments that are grounds for optimism. I am inclined to
focus on an aspect of the mind, because I believe it's the font of
our optimism. I am bullish about the mind's ability to unravel
the beliefs contained within it – including beliefs about its own
nature. What makes me optimistic is the ability of humans
everywhere to go against the grain of their beliefs – beliefs that
are familiar, feel natural and right, and once appeared to be

fundamentally true. We can (and do) unravel the contents of traditional beliefs and even the process by which they were constructed.

We've done this sort of unraveling many times before, whether it is about the relationship of the sun to the earth or the relationship of other species to us. We've put aside what seemed natural and felt right, in favor of the opposite. I am optimistic that we are now ready to do the same with questions about the nature of our own minds. From the work of pioneers such as Herb Simon, Amos Tversky, and Danny Kahneman, we know that the beliefs about our own minds that come naturally, feel right, and are easy to accept aren't necessarily true. We know that bounds on our rationality keep us from making decisions that are in our interest and that of those we love, in the long-term interest of our societies and even the planet (even perhaps the universe, with which we will surely have greater opportunity to interact in this century).

Research on the implicit beliefs and preferences we hold has shown that such 'mind bugs' extend to the beliefs and preferences we have not only about members of our own social groups but also about those farther away on the social scale. We don't intend to discriminate or treat people unfairly, but we do. Such unintended consequences come from aspects of our mind that seem natural (helping somebody close to us, like a neighbor or a nephew, rather than somebody distant) and feels right (fearing somebody who looks physically different from us). Such responses are natural and feel right because they evolved in a world where such responses were doubtless useful. And yet they continue to operate.

Becoming aware of the buggy aspects of our minds is the first step toward unraveling them. How we discover what needs unraveling, how we do it, and how successful we are at it are complex issues. But the *fact* that we do it is impressive. One of the stories to come out of the 2006 election is that some of the congressmen who had been successful at bringing home earmarks to their states lost anyway, because the voters cared about something larger than their own backyards. The ability to think about one's own long-term interest, to self-regulate, to delay gratification, to consider the well-being of the collective – especially, to view the collective as unbounded by religion, language, or nationality – requires a mental leap that isn't natural or easy, yet each new generation seems to be able to do it more successfully than the previous one. The standards for how we treat ourselves and others get higher, we examine our beliefs with more and more powerful lenses, and we turn our minds inside out and shake out the rot.

Why do we do this? For at least three reasons. First of all, because new laws demand it. Second, it is in our self-interest. Third, and most important, we unravel existing beliefs and preferences because we wish them to be in line with our intentions and aspirations and we have recognized that they are not. I see evidence of this realignment everywhere – small acts undertaken so that one may be the person one wishes to be rather than the person one is – and it is the constant attempt at this realignment that gives me optimism.

Long-Term Trends Toward Honesty to Others and Oneself

ROBERT TRIVERS

Evolutionary biologist, Rutgers University; coauthor (with Austin Burt) of *Genes in Conflict: The Biology of Selfish Genetic Elements*.

What goes up comes down, what goes around comes around, for each action there is a reaction, and so on. Life is intrinsically self-correcting at almost all its levels, including evolutionary, physiological, historical, and genetic. This permits a limited optimism. Wickedness and stupidity are ultimately self-destructive and self-limiting, so we need not trouble ourselves that any particular trend in that direction will go on indefinitely.

On the other hand, the principle of self-correction also applies to love, friendship, and high intellectual powers. No movement in these directions can proceed long without setting up counterpressures against their further spread. In short, we should neither be too despondent nor too elated at the trajectory of current events. Sooner or later – and usually sooner – they will be reversed.

Two questions arise:

Are there long-term trends we can feel optimistic about? Thirty years of work on the evolutionary trajectory of cooperative strategies suggest long-term trends (under a broad range of conditions) toward greater cooperation, contingent on ever more sophisticated discrimination. It seems likely that when similar models are produced for varying degrees of deceit and self-deception, long-term trends toward honesty to others and oneself will (at least under some conditions) be favored.

Is there any reason to believe that we will survive long enough to enjoy any of these long-term trends? This is far less certain. Evolution does not plan for contingencies that have not yet occurred – and the vast majority of species go extinct. There is no reason to expect that humans are exempt from these rules. The good news is that there is currently no chance that we could extinguish all of life. The bacterial 'slimosphere' alone extends some ten miles into the earth, and as yet we can make life truly miserable only for the vast majority of people, not extinguish human life entirely. I would expect this state of affairs to continue indefinitely. The feeling that everything may be fine if only we survive the next fifty to five hundred years may become a regular part of our psychology.

The Baby Boomers Will Soon Retire

JONATHAN HAIDT

Professor of social psychology, University of Virginia; author of *The Happiness Hypothesis: Finding Modern Truth in Ancient Wisdom*.

I am optimistic about the future of social science research, because the influence of the baby-boom generation on the culture and agenda of the social sciences will soon decrease. Don't get me wrong; many of my best friends are boomers, and technically I'm one too (born in 1963). I am grateful for the freedom and justice that the activists of the 1960s and 1970s helped bring to the United States. But if there is a sensitive period in one's lifetime for acquiring a moral and political orientation, it is one's late teens and early twenties, and most of those whose sensitive periods included the Vietnam War and the struggles for civil rights seem to have been permanently marked by those times. Many young people who entered PhD programs in the social sciences in the 1970s did so with the hope of using their research to reduce oppression and inequality. This moral imprinting of a generation of researchers may have had a few detrimental effects on the (otherwise excellent) science they produced. Here are two:

(1) *Moralistic anti-nativism.* The deep and politicized antipathy to 1970s sociobiology produced a generation of social scientists wary of nativism in general and of evolutionary thinking in particular. Nobody these days admits to believing that the mind is a blank slate at birth, but I have noticed that social scientists older than I am generally begin with a social-learning explanation of everything (especially sex differences) and then act as though it would be 'conservative' scientifically or 'liberal' politically to stick with social learning unless the evidence against it is overwhelming – it rarely is. But why privilege social learning? Shouldn't we let nativist and empiricist explanations both have a go at each question and then pick the answer that has the better fit, overall, with the evidence? I look forward to the day when most social scientists learn about the astonishing findings of twin studies in their twenties and very few know who Stephen Jay Gould was.

(2) *Moral conformity pressure.* Imagine an industry in which 90 percent of the people are men, male values and maleness

are extolled while feminine values are ridiculed, and men routinely make jokes, publicly and privately, about how dumb women are, even when women are present. Sounds like a definition of 'hostile climate' run wild? Now replace the words 'male' and 'female' with 'liberal' and 'conservative,' and we have a pretty good description of my field – social psychology – and, I suspect, many other areas of the social sciences. I have no particular fondness for conservatives. But I do have a need for them. I study morality, and I have found that conservative ideas (about authority, respect, order, loyalty, purity, and sanctity) illuminate vast territories of moral psychology – territories that have hardly been noticed by psychologists who define morality as consisting exclusively of matters of harm, rights, and justice. If social psychology had been a morally diverse field, we would have done a much better job of studying the full expanse of human morality and we'd be in a much better position right now to understand the morality of radical Islam.

Will younger social scientists be more morally diverse than the baby-boom generation? Maybe not. But if they make it through their sensitive periods without seeing themselves as part of a revolution, they just might be more open to diverse ideas about the origins of mind, the scope of morals, and the workings of society.

The Evolutionary Ability of Humankind to Do the Right Things

HAIM HARARI

Physicist, former president of the Weizmann Institute of Science.

I am optimistic about the evolutionary ability of humankind to do the right things, even though it sometimes happens only after all possible mistakes are exhausted.

I am optimistic about technology and world leaders (in that order) discovering ways to combine energy savings and alternative sources of energy (in that order), so that our planet is saved while we still have a reasonable standard of living.

I am optimistic about the irreversible trend of increasing the economic value of knowledge and decreasing the relative economic importance of raw materials, reducing the power of ruthless primitive dictators and increasing the rewards for education and talent.

I am optimistic about the emerging ability of the life sciences to use mathematics, computer science, physics, and engineering in order to understand biological mechanisms, detect and prevent medical problems, and cure deadly diseases.

I am optimistic that more scientists will understand that public awareness and public understanding of science and technology are the only weapons against ignorance, faith healers, religious fanaticism, fortune tellers, superstitions, and astrology, and that serious programs will emerge to enhance the contribution of the scientific community to this effort.

I am optimistic that, in the same way that Europe understood during the last fifty years that education for all and settling disputes peacefully are good things, while killing people just because of their nationality or religion is bad, so will the Muslim world during the new century.

I am optimistic that we will soon understand that wise medical and genetic ethics mean that we should neither absolutely forbid nor absolutely allow any technology but rather should find ways to extract the good and eliminate the bad from every new scientific development.

I am optimistic because an important fraction of the nations on this planet have succeeded in refuting the extrapolated population explosion, and I hope the remaining nations will do likewise, for their own advancement and survival.

I am optimistic about the power of education to alleviate poverty and advance health and peace in the third world, and I am hopeful that the affluent world will understand that its own survival on this planet depends on its advancement of education in the rest of the world.

I am not at all optimistic that any of the above will happen soon. All possible mistakes and wrong turns will probably be attempted. The weakest link is our chronic short-sightedness, which is bad in the case of the general public and much worse for its elected political leaders, who think in terms of months and years, not decades, and certainly not centuries.

When Men Are Involved in the Care of Their Infants, the Cultures Do Not Make War

JOHN GOTTMAN

Psychologist; founder of the Gottman Institute; coauthor (with Julie Gottman) of *And Baby Makes Three*.

In the past thirteen years, using video cameras and the Swiss Lausanne Triadic Play method, we have been able to study 222 first-born babies interacting with their new parents. I am impressed by babies, and working with them has renewed my faith in our species. However, in the first study we did, with 130 newlywed couples, we discovered the grim fact that 67 percent of couples experienced a large drop in relationship satisfaction in the first three years of their baby's life. We also found that hostility between parents increased dramatically. The baby was deeply negatively affected by this increased hostility. In fact, from the way a couple argued in the last trimester of pregnancy, we could predict with high accuracy how much their baby would cry.

But then we compared the 67 percent with the 33 percent of couples who did not experience that negative drop in happiness when their first baby arrived, and the two groups turned out to be very different even a few months after the wedding. So my wife and I designed an educational workshop based on these differences. What I am really optimistic about is that now we have discovered, in two randomized clinical trials, that in just a two-day workshop we can reverse the negative effects of the arrival of a first baby. Furthermore, there is a dramatic change in the fathers, and this has a marked effect on the emotional and neurological development of their babies (even though the babies didn't take the workshop).

The other thing I'm optimistic about is how much men have changed in the past thirty years. Thirty years ago, we'd have only women in our audiences. Men who are becoming dads want to attend these workshops; they want to be better partners and better fathers than their own dads were. We have found that change in all walks of life, all socioeconomic levels, all the races and ethnic groups we've worked with in this country. We have now trained workshop leaders in twenty-four countries. I believe that these workshops can change families, avoid the deterioration of couples' relationships, and contribute to what Dan Goleman has called 'social intelligence' in a new generation of children. Anthropologist Peggy Sanday's study of 186 hunter-gatherer cultures found that when men are involved in the care of their infants, the cultures do not make war. That greater involvement of men with their babies may eventually contribute to a more peaceful world. That thought makes me optimistic.

The Survival of Friendship

JUDITH RICH HARRIS

Independent investigator and theoretician; author of *No Two Alike: Human Nature and Human Individuality*.

I am optimistic about human relationships – in particular, about friendship. Perhaps you have heard gloomy predictions about friendship: It's dying out, people no longer have friends they can confide in, loneliness is on the rise.

But friendship isn't dying out, it's just changing, adapting to the changes in the world. People are discovering different ways of getting together. It may be harder to find a bowling partner, but it's easier to find someone to chat with, because there are many more ways to chat.

When I was a child, people with chronic illnesses were described as 'shut-ins.' Now a person can be shut in without being shut out. I have friends whom I know only through e-mail conversations but who are as dear to me as my college roommate and dearer by far than my next-door neighbor.

The desire to form and maintain relationships is one of the built-ins of human nature. Primates are social animals, and humans are the most social of all. An extravagant amount of mental capacity is devoted to relationships. We can recognize at a glance the faces of thousands of different people and, with equal ease, remember whether or not we like them. With a bit more effort, we can dredge up other useful information about most of them: their names or professions or where we met them. Throughout our lives, we collect and store information about specific individuals, so that – just in case we ever run into them again – we will know how to act. We even store information about people we have never met and whose faces we have never seen.

Collecting people information is something we do without training and with no reward other than the enjoyment we get from doing it. We don't need a nudge from the conscious mind telling us that the information may come in handy someday. But in fact it may come in handy. People we have never met before may be important to us in the future. They may become our trading partners or employers. They may become our lovers or our rivals.

Or they may simply become our friends.

The Public Will Become Immune to Hype

ROGER HIGHFIELD

Science editor, *The Daily Telegraph*; coauthor (with Ian Wilmut) of *After Dolly: The Uses and Misuses of Human Cloning*.

I am quietly optimistic that in the wake of years of hype over the practical significance of gene discoveries, fusion power, 'magic bullets,' superconductivity, gene therapy, cures for aging, and embryonic stem cells,* the public will become more pessimistic about the practical benefits of discoveries made in the lab and more appreciative of what science is really about – basic curiosity, rationality, and the never-ending dialog between ideas and experiments. With luck, the public will spend more time gazing up at the blue skies of science and not down at the brown torrent of parochial and humdrum expectations about what science can do for them. Science does not have to be useful, save to put forward useful models of how nature works. Science does not have to cure disease. Science does not have to make us live to the age of a hundred and twenty. Science does not have to make money.

* Being a science journalist, I plead guilty on all counts.

Solving the Mind-Body Problem

DONALD D. HOFFMAN

Professor of cognitive science and of information and computer science, University of California, Irvine; author of *Visual Intelligence*.

The enigmatic relation between conscious experiences and the physical world, commonly known as the mind-body problem, has frustrated philosophers at least since Plato and now stonewalls scientists in their attempts to construct a rigorous theory. Yet I am optimistic that, despite millennia of failures, we will soon devise a scientific theory for this perennial problem.

Why such optimism? First, the mind-body problem is now recognized as a legitimate scientific problem. In 2005, the journal *Science* placed it second in a list of 125 open questions in science. During the 20th century, a multi-decade detour into behaviorism sidelined scientific investigation of the mind-body problem. But three decades into the cognitive revolution, the problem was dusted off and again given serious scientific attention.

Second, scientists soon rediscovered that the problem is surprisingly hard. Neurophysics, real and artificial neural networks, classical and quantum algorithms, information and complexity – standard tools that prove powerful in the study of perception, cognition, and intelligence – have yet to yield a single scientific theory of conscious experiences. We cannot, for instance, answer the basic question: Why must *this* particular pattern of neural activity, or *this* functional property, cause or be *this* particular conscious experience (say, the smell of garlic) instead of *that* conscious experience (say, the smell of a truffle), or instead of no conscious experience at all? Precise predictions of this type, *de rigueur* for genuine scientific explanations, have yet to be fashioned, or even plausibly sketched, with the standard tools.

Third, although science is laudably conservative, scientists, when pushed to the wall by recalcitrant data and impotent theories, have repeatedly proved willing to reexamine dearly held presuppositions and to revise or jettison the ineffectual in favor of unorthodox assumptions – provided those assumptions permit the construction of explanatory theories that conform to the data. Aristarchus, then Copernicus, proposed a heliocentric solar system; Newton proposed action at a distance; Einstein proposed quanta of light and distortions of spacetime; Bohr proposed probability waves, superpositions, and nonlocality. Theories of quantum gravity now posit at least eleven dimensions, vibrating membranes, pixels of space and time. The initial response to such proposals is invariably widespread incredulity, but experimental confirmation – or considerations of explanatory power, wherever they point – eventually win the day. Scientists revise their offending presuppositions, adjust as best they can to the new worldview, and get on with the business of science in the new framework.

Evidence is mounting that the mind-body problem requires revision of deeply held presuppositions. The most compelling evidence to date is the large and growing set of proposals now on offer, all of which are nonstarters. They are (to quote Wolfgang Pauli) not even wrong. We have yet to see our first genuine scientific theory of the mind-body problem. This has prompted some to conclude that *Homo sapiens* has been cheated by evolution and simply lacks the requisite concepts: Those concepts necessary for us to survive long enough to reproduce did not include those necessary to solve the mind-body problem. If so, there is little hope (at present) for swift progress.

I am optimistic, however, that the obstacle is not in our genes but in our presuppositions. Tinkering with presuppositions is more clearly within the purview of current technology than tinkering with our genes. Indeed, tinkering with presuppositions requires no technology at all, just a ruthless reconsideration of what one considers to be obviously true. Science has risen to that task before. It will do so again. But progress will be tortuous and the process psychologically wrenching. It is not easy, even in the light of compelling data and theories, to let go of what once seemed obviously true.

Here are some obvious truths that guide current attempts to solve the mind-body problem: Physical objects have causal powers. Neural activity can cause conscious experiences. The brain exists whether or not it is observed. So too does the moon, and all other physical objects. Consciousness is a relative latecomer in the evolution of the universe. Conscious sensory experiences resemble, or approximate, true properties of an independently existing physical world.

Will we soon be forced to relinquish some of these truths? Probably. If so, the current ontological predilections of science will require dramatic revision. Could science survive? Of course. The fundamental commitments of science are methodological, not ontological. What is essential is the process of constructing rigorous explanatory theories, testing them with careful experiments, and revising them in light of new data. Ontologies can come and go. A particular one might endure long enough to be taken for a *sine qua non* of science. But it is not: An ontology given life by the method of science can later be slain by that same method. Therein lies the novel power of science. And therein lies my optimism that science will soon succeed in fashioning its first theory of the mind-body problem. But at the feet of that theory will probably lie the corpse of a slain ontology.

Print As a Technology

WALTER ISAACSON

President and CEO of the Aspen Institute; former CEO of CNN and managing editor of *Time*; author of *Benjamin Franklin: An American Life*.

I am very optimistic about print as a technology. Words on paper are a wonderful information-storage, retrieval, distribution, and consumer product. (That is why I appreciate the fact that many *Edge* forums are transformed into books; I hope someday there will be a gorgeous *Edge Magazine* to flip through.) Imagine if we had been getting our information delivered digitally to our screens for the past 550 years. Then some modern Gutenberg had come up with a technology that was able to transfer these words and pictures onto pages that could be delivered to our doorstep, and we could take them to the backyard, the bath, or the bus. We would be thrilled with this technological leap forward, and we would predict that someday it might replace the Internet.

Truth Prevails – Sometimes Technology Helps

XENI JARDIN

Tech-culture journalist; coeditor of *BoingBoing*; NPR commentator; columnist for *Wired*.

I became a born-again optimist this year in an unlikely place: surrounded by hundreds of cardboard boxes filled with the dead.

They were victims of Guatemala's civil war. Row after row, stacked floor to roof, on the top level of a building protected by concertina wire and armed guards in Guatemala City. This site is home to a group called the Forensic Anthropology Foundation of Guatemala.

The living were downstairs. Using open-source software, recycled computers, and DNA forensics help from labs in the United States, they work to identify the dead. The FAFG staff includes lawyers, *antropologos forensicos*, and IT engineers. The process begins when someone tips them on the whereabouts of one of the clandestine mass graves. Then comes the slow digging and what must be painful conversations with the surviving relatives, who often fear retribution from the perpetrators (the killers sometimes live in the same village, maybe right next door). The army recruited soldiers from the same Mayan villages their scorched-earth policies sought to destroy.

The FAFG exhumations yield clumps of bones, flesh, sometimes the clothing the victim wore when the killing happened. And back in this Guatemala City building now, the living are cleaning and scraping and sorting those clumps of bone and dirt, laying them out on tables, brushing the soil off, marking each tibia and fibula and tooth with codes that will soon be tapped into databases. When everything comes together just right – the survivors' testimony, the database tables, the DNA prints, the bullet holes through the dry cranium, the dig maps – when all of that clicks, someone then writes a code in black marker on the side of a cardboard box: 'Jacinto Rodriguez: FAFG-482-V-I, Nebaj, El Quiché.' He was one of thousands whose deaths the military authorities denied or disavowed for decades.

Sometimes, governments turn on their own citizens, and those corrupt regimes are sustained in part by lies. Sometimes the lies last for decades. Sometimes longer. But science does not lie.

These boxes full of bones, and all the data with which they're tagged – none of that lies. Even though the living in this building work under death threats (they're texted in by SMS now), even though they lack financial, technical, and practical resources, more and more of those boxes, day after day, fill with codes and names. And eventually the dead return to their pueblos, inside these boxes, for reburial.

'The survivors want to know that their family members will rest in a dignified way, instead of being dumped by the side of the road like dogs,' one of the anthropologists told me. 'More than justice in the American sense of the word – more than revenge, or legal process – they just want their people back.'

I have recently met with other organizations like the FAFG, in Guatemala and other countries – organizations run by individuals who are working very hard, under impossibly difficult conditions, to uncover and preserve the truth of past human-rights violations. And what I've seen – in particular, new uses of technological tools to solve old problems – has given me hope. The truth eventually prevails. While at least one person believes that the truth matters, there is hope.

Human Intelligence Can Be Increased Dramatically

STEPHEN M. KOSSLYN

John Lindsley Professor of Psychology, Harvard University; coauthor (with Olivier Koenig) of *Wet Mind: The New Cognitive Neuroscience*.

I am optimistic that human intelligence can be increased dramatically in the near future. I see three avenues that will lead to this end.

First, cognitive neuroscience and related fields have identified a host of distinct neural systems in the human brain. Various combinations of these systems are used in accomplishing different tasks, and each system can be made more efficient by targeted training. Such training involves having people perform tasks designed to exercise specific abilities, which grow out of distinct neural networks. Just as a body builder can do curls to build up biceps and dips on parallel bars to build up triceps, we can design computer-game-like tasks that exercise specific parts of the brain – mental muscles, if you will. By exercising the right sets of systems, specific types of reasoning not only can be improved but (the holy grail of training studies) such improvement can generalize to new tasks that draw on those systems.

Second, people often grapple with problems in groups, be they formally designated teams or casual huddles around the water cooler. I am optimistic that understanding the nature of such group interactions will increase human intelligence. Just as a mechanical calculator can extend our mental capacities, other people help us extend our intelligence – both in a cognitive sense (as required to solve problems) and in an emotional sense (as required to detect and respond appropriately to emotions, ours and those of others). In this sense, other people can serve as social prosthetic systems, extensions of our own brains. Just as a wooden leg can fill in for a missing limb, others' brains can fill in for our own cognitive and emotional limitations. To the extent that researchers come to understand how such social prosthetic systems arise and operate, they will understand how to increase human intelligence.

Third, the line between animate and inanimate information-processing is becoming increasingly blurry as research in multiple fields proceeds apace. I expect that engineers will continue to press forward, designing increasingly powerful machines to help us extend our intelligence. For example, some people carry computers with them everywhere they go and treat Google as an extension of their own knowledge bases. Or, in my case, my PDA extends my organizational ability enormously. We soon will have a wide variety of mechanical helpmates. The distinction between what goes on in the head and what relies on external devices is becoming more subtle and nuanced, and in this way human intelligence is being extended.

Crucially, each of these three developments amplifies the effects of the others, producing synergies: Just as brain exercises enhance our intellectual abilities, we can learn how to make better use of mechanical aids and how to rely more effectively on other people. The confluence of all three developments will produce positive-feedback loops, whereby the very act of interacting with others or working with smart devices will help us continue to develop our brains, and as our brains develop we will in turn be able to use increasingly sophisticated devices and rely on people in more complex and powerful ways.

With luck, such developments will produce new sorts of extended social links and highly integrated social networks, and a new kind of 'smart society' will emerge. And who knows? Such a society may be not only smarter but also wiser.

A New Contentism

KAI KRAUSE

Software and design pioneer.

It's a turquoise sky on a December afternoon. Out the window, I see the Rhine meandering far below me, the last rays of sunlight shimmering on the surface; a storm is passing through from France heading toward Cologne. To check on it, I push a button, the screen fills with dozens of little widgets: I have a quick look at the real-time weather animation loop for the last six hours and can see that most of it will pass to the north.

A little bell sound – ah, an SMS text message coming in from my daughter in London: her final cake worked well! I smile – bet she is happy now, on the way to her diploma in *patisserie* at the Cordon Bleu cooking school. I flip open the cell phone qwerty keyboard and type a smiley with hugs to her with both thumbs.

Then a window pops up on the big screen: It's my old pal Ben, who's on-line in Santa Barbara. In the chat window, he sends me an image from NASA: new evidence for water flowing down the sides of a crater on Mars. He also pastes in a link to the NASA site, the superresolution files. I download a handful and watch a couple of hundred megabytes come down.

That's a case for the other machine: I drag the images over to the T221, 9 million pixels at 200dpi – what a wonder to behold! Water on Mars! We had just seen the BBC documentary on the amazing tenacity of life, the Mariana Trench, 11,000 meters below the surface – they keep finding life of surprising complexity in the most unlikely places. I smile again: In my lifetime, there may be irrefutable evidence of life beyond Earth.

Ben suggests a Go game, but now there is a window with Matt and Mike. We have been working on cool new software for quite a while, and there is one twist: They live outside Auckland, in the forests of northern New Zealand, the antipode of my place in Europe. But it works well: The new version is ready. Matt sends it encrypted, while the three of us chat, with annotated screenshots of bugs. We could switch to VoiP or live camera, but it's helpful to have the written record as well.

I yank the mouse to a corner. The widgets reappear and a few world-time clocks tell me that it's 8:00 am where they are – Wow, up early! I glance at some headlines on slashdot, digg, heise, arstechnica, perlentaucher. I see the NPR news headlines, the Tate Gallery RSS shows an exhibition of Holbein – sweet. Didn't I just read about him on edge.org? Where is that Tate St. Ives anyway? I switch to Google Earth and swoop down on the U.K. from space.

I drag the mouse to the other corner. The screen fills with tiny colorful images, probably a hundred of them, all the files currently open shown in miniature. I find the NASA shots and drag them over to the side into a folder for science stuff. Quite a collection of bits and pieces there, years of collecting them. Attached for a few hundred bucks is a terabyte's worth of muse and ponder. Dozens of reference books, the *Encyclopaedia Britannica*, what a dream!

Come to think of it, I should send that to my boys, who are eleven and fifteen, a voracious age. I ping the younger one, ask if he has time for Dad just now. I send him a sweet physics simulation, *LineRider,* and the lovely *SandGame* particle system applet; he will have fun with that. In the meantime, a quick scan of the news for a good description of the water-on-Mars events. *New Scientist* has a nice write-up, *Wired* a good story with side-by-side comparisons. I print the page into a pdf file, drag it over to him. The older one pops up, has some new jokes for me. He loves to read them and to create his own. I tell him to google '2.3 trillion missing' and watch the YouTube video. Ponder a little.

They will come over here soon. We can go to the medieval Christmas market. Drive down to Innsbruck for real snow. I should get a few movies for us to watch. Are they ready for *Aguirre?* Maybe *2001?* Christopher sent *The Fifth Element* on Blu-ray. Perhaps *Amélie.* Next summer we go to Paris. Now that piano is in my ear. A quick change to the music store, a search for Yann Tiersen; there is almost everything he ever made. A few clicks, a few bucks, and the soundtrack is here. I burn a CD for the trip. Click in Wikipedia; what has he been up to? Duet with Jane Birkin and Cocteau Twins – neat, missed that!

My much better half comes over for a hug. Time to go, we have lots of errands before the holidays. She flies along the *Autobahn* at over 220 km/h with a grin; the navigator detects a traffic jam ahead, rerouting us across the river – just a ten-minute detour, not bad. Could get some gas though: All the nearest gas stations are shown. I click on one, it takes us there. Pause the favorite song while I go in, grab a few magazines and newspapers. No cash; I slide the plastic through the slit and leave. She had already entered the address we want to go to first – tricky access through the maze of one-way streets around the old-town area. Minus 3° outside, we snuggle in the heated seats, on the screen the closest Thai restaurants . . .

Obviously I could go on *ad nauseam* here, but this is not a description of technology per se. The emphasis is on quality of life. On the benefits of tools, the liberating freedom. My real point: Humans are feeble. We forget. We have become numb to all the wonder.

To see the weather in pictures from space, animated over time – what a wonder that would have been to the Wright brothers, or James Cook, or Vasco da Gama, or Marco Polo. To be in realtime communication with your family – what a wonder that would have been for Bach who had twenty children. (Half died in infancy. I haven't even touched on the advances in health and medicine.)

To see cell phones and billions of SMS would have astonished Tesla, Edison, Bell, Reis, Meucci. To personally own the images from a planetary probe in startling clarity – what a dream that would have been for a Huygens, Mercator, Kepler, Galileo!

To collaborate on your work with colleagues on the other side of the world as if they were in the next room – how liberating is that freedom! To travel safely, quickly, effortlessly, with an all-knowing friend guiding you – what would any of them say to that? Researchers have calculated that Goethe traveled over 37,000 kilometers in his lifetime in more than 180 excursions – on foot, horseback, and carriage. Add another zero for a guy like Humboldt. They would have marveled – or cried – at our ability to go anywhere, see anything, meet anyone.

To be able to see all the works of all the great artists and keep a copy to examine up close, at your leisure, in your own home. To listen to the music of any composer, new or old – what an absolute dream that would have been, for any and all of them! Imagine that you'd heard about that new Beethoven symphony: You would have had to physically travel to a performance somewhere, and even then you could have heard only that one performance, not any of the others, and you would likely forget the music, since you would hardly ever get a chance to hear it again, to build a long-term memory of it. Never mind mentioning movies here, or radio and television, let alone the Web.

To get to research done anywhere, by anyone; to share the findings and writings; duplicate them instantly, store them and save them; catalog them and index them, searchable among billions, in seconds! To have your own copy of the books, your own *Britannica* – how blissful that would have made Jules Verne, with his 20,000 wooden boxes of index-card snippets, or any of the other polymaths, like Athanasius Kircher, or T.H. Huxley, or Newton, or Leibniz. To have your own diviner of answers to any question, finder of any fact, in minutes or even seconds, an advisor like no Sun King or emperor, kaiser or pharaoh, could ever buy with all the gold in his empire. That's Google, now, in a smartphone, in the pockets of teenagers.

Surely any of the ancient Greeks would spontaneously combust at the sight of almost any household object on any shelf in any department store. And yes, I realize that that won't make us automatically and constantly happy. Is there a list of evils and downsides that came with all those advances? You bet! I could switch hats and drown in drivel about all that.

But the question stands: What am I optimistic about? And I think it bears repeating: Countless scientists over the millennia dedicated their lives to discoveries, to solutions, to inventions and explanations. They had visions of bettering the fate of humanity, of seeking truths and finding answers, and they paid for it with enormous efforts and in many cases with their lives. Their combined body of bodies stands in front of us, in awe and *in tears!* We have achieved almost all their dreams. We have freedom, in every sense, as never before in history, and *we are ungrateful bastards about it!*

Let us just be content again. Plain happy. Period. I am calling for a New Contentism.

From that vantage point, looking at the incredible options and tools for all of us, is there reason to be optimistic that in the future we might make good use of them? You bet!

The Young Will Take Repair of the World into Their Own Hands

HOWARD RHEINGOLD

Communications expert; author of *Smart Mobs: The Next Social Revolution.*

The tools for cultural production and distribution are in the pockets of fourteen-year-olds. This does not guarantee that they will do the hard work of democratic self-governance: The tools that enable the free circulation of information and communication of opinion are necessary but not sufficient for the formation of public opinion. Ask yourself this question: Which kind of population seems more likely to become actively engaged in civic affairs – a population of passive consumers sitting slack-jawed in their darkened rooms, soaking in mass-manufactured culture broadcast by a few to an audience of many, or a world of creators who might be misinformed or ill-intentioned but are nevertheless actively engaged in producing, as well as consuming, cultural products? Recent polls indicate that a majority of today's youth – the digital natives for whom laptops and wireless Internet connections are part of the environment, like electricity and running water – create as well as consume on-line content. This bodes well for the possibility that they will take the repair of the world into their own hands instead of turning away from civic issues or toward nihilistic destruction.

The eager adoption of Web publishing, digital video production and on-line video distribution, social networking services, instant messaging, multiplayer role-playing games, on-line communities, virtual worlds, and other Internet-based media by millions of young people around the world demonstrates the strength of their desire – unprompted by adults – to learn digital production and communication skills. Whatever else might be said of teenage bloggers, dorm-room video producers, or the millions who maintain pages on social networks like MySpace and Facebook, it cannot be said that they are passive media consumers. They seek, adopt, appropriate, and invent ways to participate in cultural production. While moral panics concentrate the attention of oldsters on lurid fantasies of sexual predation, young people are creating and mobilizing politically active publics on-line when circumstances arouse them to action. Last year, 25,000 Los Angeles high school students used MySpace to organize a walk-out from classes to join street demonstrations protesting proposed immigration legislation. Other young people have learned how to use the sophisticated graphic-rendering engines of video games as tools for creating their own narratives. In France, disaffected youth, the ones whose riots are televised around the world but whose voices are rarely heard, used this emerging 'machinima' medium to create their own version of the events that had triggered their anger. (Search for 'The French Democracy' on video-hosting sites). Not every popular YouTube video is a teenage girl in her room (or a bogus teenage girl in her room); increasingly, do-it-yourself video has been used to capture and broadcast police misconduct or express political opinions. Many of the activists who use Indymedia (ad-hoc alternative media organized around political demonstrations) are young.

My optimism about the potential of the generation of digital natives is neither technological determinism nor naïve utopianism. Many-to-many communication enables but does not compel or guarantee widespread civic engagement by populations who never before had a chance to express their public voices. And while the grimmest lesson of the 20th century is to mistrust absolutist utopians, I perceive the problem to be in the absolutism more than the utopianism. Those who argued for the abolition of the age-old practice of human slavery were utopians.

Toward a Broader Sense of Global Issues and Possibilities

LINDA STONE

Former vice-president of Microsoft and cofounder and former director of its Virtual Worlds Group.

Ten years ago, the novelist Andrei Codrescu came to visit me at Microsoft, where I was director of the Virtual Worlds (now Social Computing) Group. As he watched me engage in conversation in V-Chat and Comic Chat, he mused skeptically about these virtual communities – until a soldier entered the conversation.

'Ask him where he is,' Codrescu demanded. The soldier's reply, 'Stationed in Germany, fighting in Bosnia.' Andrei grabbed the keyboard from me, full attention now on the soldier, as he was sucked into the virtual world. When he finally disengaged, he seemed fascinated by its possibilities.

Technology has advanced so significantly in the last decade, and many in the generation of kids high school age and younger are so fluent in every aspect of it that they have moved beyond being participants stimulated by the technology to being creators of both technology and content, collaborating and sharing every aspect of their lives, their opinions, and their causes across borders that, to them, are increasingly invisible.

Anything can change when we change the way we look at it, and the generation coming of age sees a more global world and experiences a range of resources for creation and collaboration on a scale previous generations could only imagine. Through podcasts, YouTube, blogs, MySpace, and emerging technologies, every issue we face today – from successful alternative-energy solutions to avian-flu outbreak areas to disaster recovery – is part of the global conversation, and there are many pathways to participate and co-create solutions.

Attention is the most powerful tool of the human spirit. The opportunity this generation appears to be seizing is to move our collective attention away from the anxiety and despair fueled by campaigns like the War on Terror and toward the most positive future we can create together – as highlighted by blogs like Worldchanging.com.

Through shared experiences, the generation now growing up has a broader sense of global issues and possibilities and a reality that moves beyond 'my country' and instead embraces 'our planet.'

Optimism on the Continuum Between Confidence and Hope

RAY KURZWEIL

Inventor and technologist; author of *The Singularity Is Near: When Humans Transcend Biology*.

Optimism exists on a continuum in between confidence and hope. Let me take these in order.

I am *confident* that the acceleration and expanding purview of information technology will solve within twenty years the problems that now preoccupy us.

Consider energy. We are awash in energy (10,000 times more than required to meet all our needs falls on Earth) but we are not very good at capturing it. That will change with the full nanotechnology-based assembly of macro objects at the nano scale, controlled by massively parallel information processes, which will be feasible within twenty years. Even though our energy needs are projected to triple within that time, we'll capture that .0003 of the sunlight needed to meet our energy needs with no use of fossil fuels, using extremely inexpensive, highly efficient, lightweight, nano-engineered solar panels, and we'll store the energy in highly distributed (and therefore safe) nanotechnology-based fuel cells. Solar power is now providing 1 part in 1,000 of our needs, but that percentage is doubling every two years, which means multiplying by 1,000 in twenty years. Almost all the discussions I've seen about energy and its consequences (such as global warming) fail to consider the ability of future nanotechnology-based solutions to solve this problem. This development will be motivated not just by concern for the environment but also by the $2 trillion we spend annually on energy. This is already a major area of venture funding.

Consider health. As of just recently, we have the tools to reprogram biology. This is also at an early stage but is progressing through the same exponential growth of information technology, which we see in every aspect of biological progress. The amount of genetic data we have sequenced has doubled every year, and the price per base pair has come down commensurately. The first genome cost a billion dollars. The National Institutes of Health is now starting a project to collect a million genomes at $1,000 apiece. We can turn genes off with RNA interference, add new genes (to adults) with new reliable forms of gene therapy, and turn on and off proteins and enzymes at critical stages of disease progression. We are gaining the means to model, simulate, and reprogram disease and aging processes as information processes. In ten years, these technologies will be 1,000 times more powerful than they are today, and it will be a very different world, in terms of our ability to turn off disease and aging.

Consider prosperity. The 50-percent deflation rate inherent in information technology and its growing purview is causing the decline of poverty. The poverty rate in Asia, according to the World Bank, declined by 50 percent over the past ten years due to information technology and will decline at current rates by 90 percent in the next ten years. All areas of the world are affected, including Africa, which is now undergoing a rapid invasion of the Internet. Even sub-Saharan Africa has had an average annual 5-percent economic growth rate in the last few years.

OK, so what am I *optimistic* (but not necessarily confident) about?

All of these technologies have existential downsides. We are already living with enough thermonuclear weapons to destroy all mammalian life on this planet – weapons that are still on a hair-trigger. Remember these? They're still there, and they represent an existential threat.

We have a new existential threat, which is the ability of a destructively minded group or individual to reprogram a biological virus to be more deadly, more communicable, or (most daunting of all) more stealthy (that is, having a longer incubation period, so that the early spread is undetected). The good news is that we have the tools to set up a rapid-response system like the one we have for software viruses. It took us five years to sequence HIV, but we can now sequence a virus in a day or two. RNA interference can turn viruses off, since viruses are genes, albeit pathological ones. Sun Microsystems founder Bill Joy and I have proposed setting up a rapid-response system that could detect a new virus, sequence it, design an RNAi (RNA-mediated interference) medication, or a safe antigen-based vaccine, and gear up production in a matter of days. The methods exist, but as yet a working rapid-response system does not. We need to put one in place quickly.

So I'm optimistic that we will make it through without suffering an existential catastrophe. It would be helpful if we gave the two aforementioned existential threats a higher priority.

And, finally, what am I *hopeful*, but not necessarily optimistic, about?

Who would have thought right after September 11, 2001, that we would go five years without another destructive incident at that or greater scale? That seemed unlikely at the time, but despite all the subsequent turmoil in the world, it has happened. I am hopeful that this respite will continue.

Skeuomorphism

TIMOTHY TAYLOR

Archeologist, University of Bradford, U.K.; author of *The Buried Soul: How Humans Invented Death*.

I am optimistic about skeuomorphism. Odd, perhaps, but true.

In a small wire tidy on my desk, I have several corks. But they are not cork. The word 'cork' comes from the Latin for oak – *quercus*, a subspecies of which has the spongy bark so useful for sealing wine in bottles. In the 1980s, demand for high-quality cork began to outstrip supply. As low-grade cork often taints (or 'corks') wine, substitutes were sought. My corks are synthetic. One is cork-colored and slightly variegated, to make it appear traditional; like real corks in the German Riesling tradition, it is stamped in black with a vine-tendril motif. Another is less convincingly mottled and is mid-yellow in color, with the name of the vintner, Gianni Vescovo, printed in bold black. Both these corks are skeuomorphs – objects that preserve formal vestiges of the constraints of an original which are no longer strictly necessary in the new material. First-generation skeuomorphs are close mimics, even fakes.

Second-generation skeuomorphs, like the Vescovo cork, abandon any serious attempt at deception. Its mottling, and the fact that it is still a functional cork rather than a metal screw-top closure (equally efficient for the modest young wine it briefly protected), is a comforting nod to the history of wine. At the same time, it signals a new, more consistent freedom from contamination.

As synthetic corks became more familiar, new and more

baroque forms arose. These third-generation skeuomorphs are fun: a bright purple cork that stoppered an Australian red suggests a grape color, while a black cork has a high-tech look that draws symbolic attention to the new techniques of low-temperature fermentation in stainless steel. This black cork is still mottled, but in an exaggerated and unconvincing manner – a self-conscious and playful back-reference both to real corks and to earlier skeuomorphic examples. One could not conceive of the black cork without a process of skeuomorphic familiarization, through first- and second-generation examples. Put the black cork next in sequence with a real cork and the dissonance would be too great.

I see much of the history of technology as an unplanned trajectory in which emergent skeuomorphic qualities often turn out to have been critical. Corks are a relatively trivial example in an extraordinary history of skeuomorphism impossible to review here but which encompasses critical turns in material development – from prehistoric flint, through the discovery of metals and alloys, to complex compound objects, of which computers are a modern manifestation.

My optimism about skeuomorphs arises, as optimism often does, from former pessimism. I grew up with Alan Turing's unsettling vision of a future machine indistinguishable from a human in its reactions. Ray Kurzweil's provocative prediction of the impending 'singularity' – the point when computer intelligence would start to leave humans gasping in its intellectual wake – added to my fears. I actually began to worry that efforts to enculture my children with Shakespeare and Darwin, and even with spiritual and moral values, might be rendered peremptorily redundant by cold robotic *Übermenschen*.

I have recently become quite relaxed about all this, but not because I doubt for a moment that computers are rapidly becoming very smart indeed and will become smarter, in some directions, than we can easily imagine. Computers explicitly reproduce aspects of the human brain. Yet their eventual power will probably not be in simulation or deception. There will never be a perfect Turing machine, except under conditions so artificial and in contexts so circumscribed as to be rather ridiculous. Instead, by surpassing us in some areas, computers will relieve our brains and bodies of repetitive effort. But it will not be mimicry of our brain function that will be important. If they behave as other skeuomorphs before them, it will be computers' currently unimagined emergent qualities that we will come to value most, enhancing and complementing our humanity rather than competing with and superseding it.

In like fashion, the synthetic corks have taken the pressure off the oak groves, securing their future and with it those genuine Champagne moments. Happy New Year!

The Rise of Usability

MARTI HEARST

Computer scientist, School of Information, University of California, Berkeley.

I am optimistic that the values and principles of good user interface design will increasingly be practiced by technology designers and expected by technology users.

The design of a computer user interface can influence whether the outcome of an election is fair, a patient receives the right medicine, or a helicopter pilot makes a safe landing. Perhaps more important, when technology is designed to mesh well with how people think and how they want to live, it enhances and enriches their lives. Poorly designed interfaces that treat us inhumanely can contribute to the feelings of apprehension we often associate with our technology-saturated world.

Good interface design is currently more of a practice than a science, although the field of human/computer interaction makes use of results from cognitive science and relevant branches of psychology where applicable. The rise of the World Wide Web has accelerated advances in an understanding of usability, chiefly for two reasons. First, before the Web, most complex computer interfaces were visible only to those who could afford expensive software packages; most users would not see more than one design for a given application. The Web, in yet another example of its leveling effect, allows nearly everyone to see nearly every interface. Thus designers can learn rapidly from what others have done, and users can see if one Web site's experience is substandard compared with others. Second, the default setting for Web pages and browsers allows designers to see the source code behind each Web page and thus learn from one another how to code the best designs. The rise of the Web has increased consumer demand and awareness of good design and has increased the value of user interface design within computer science departments.

Unfortunately, misguided views about usability still cause significant damage in today's world. In the 2000 national election, poor ballot design led thousands of voters in Palm Beach,

Florida, to vote for the wrong candidate, thus turning the tide of the entire presidential election. At the time, some observers made the ignorant claim that voters who could not understand the Palm Beach butterfly ballot were not bright enough to vote. I wonder whether the people who made such claims have never made the mistake of, say, trying to pull open a door that requires pushing. Usability experts see this kind of problem as an error in the design of the door rather than a problem with the person trying to open it.

It appears that some Florida election officials did not learn the lessons of the 2000 election. Evidence suggests that in Sarasota County poor design of an electronic ballot led to significant undervoting in the 2006 mid-term elections. One in seven voters did not mark any choice in a hotly contested congressional race that was decided by fewer than 400 votes. Usability mistakes similar to those in Palm Beach were repeated in Sarasota. But it is unfair to expect election officials to be experts in usability and information design; rather, what is lacking is a general recognition that an understanding of how people think and use technology is integral to ballot design and fair elections. Just as computer experts design and maintain the electronic voting machines, usability experts should assess all ballot designs before they're released to the public. In the future, we may well see automated usability checkers (akin to grammar checkers in word processors) for applications such as ballot design.

Ballot design is just one example for which interface design matters. There is ample evidence that good design reduces errors and increases enjoyment of the use of technology. Fortunately, an appreciation of the importance of good design for human use of technology is growing.

Interpersonal Communication Will Become More Profound; Rationality Will Become More Romantic

JARON LANIER

Computer scientist, virtual reality pioneer, and musician.

The future may be transformed by the very nature of communication, much as it was when language appeared. This is not easy to imagine, but here's one approach to thinking about it: I've been fascinated by the potential for postsymbolic communication for years. This new style of interpersonal connection could become possible, once large numbers of people become virtuosos at improvising what goes on in virtual reality.

We are virtuosos at spoken language. Adults speak with what seems like no effort at all, even though everyday chats might be the most complicated phenomena ever observed. I see no reason why new virtuosities in communication could not appear in the future, though it's hard to specify a timeframe.

Suppose you're enjoying an advanced future implementation of virtual reality, and you can cause spontaneously designed things to appear and interact with the ease of sentences pouring forth in an ordinary conversation. Whether this is accomplished by measuring what your body does from the outside or by interacting via interior states of your brain is nothing more than an instrumentation question. Either way, we already have some clues about how the human organism might be able to improvise the content of a virtual world.

Some of the most interesting data from VR research thus far involve homuncular flexibility. It turns out that the human brain can learn to control radically different kinds of bodies with remarkable ease. Thus people might eventually learn to spontaneously change what's going on in a virtual world by becoming parts of it.

That aspect of the brain optimized to control the many degrees of freedom of body motion is also well suited to controlling the many degrees of freedom of a superlative programming and design environment of the future. (It is likely, by the way, that the tongue would turn out to be just as important in this type of communication as it is in language, for it is the richest output device of the human body.) Why bother? It's a reasonable hunch. Words have done so much for people that it's reasonable to suppose that alternatives to them, with overlapping but distinct functions, will lead to new ways of thinking and connecting. An alternative to abstraction might arise: the possibility of expression through a fluid and capable concreteness. Instead of the word 'house,' you could conjure up a particular house.

How do you know it's a house without using the word? Instead of falling back on whatever the word 'house' means, you might toss around a virtual bucket that turns out to be very large on the inside and contains a multitude of house prototypes. In one sense, this 'fuzzy' collection is more precise than the word – in another, less so.

If all this sounds a little too fantastic or obscure, here's another approach to the same idea using more familiar reference points. Imagine a means of expression that is a cross between the three great new art forms of the 20th century: jazz improvisation, computer programming, and cinema. Suppose you could improvise anything that could be seen in a movie with the speed and facility of a jazz improviser. What would that mean for the sense of connection between you and someone you love?

There's a little book by James P. Carse with a wonderful title, *Finite and Infinite Games*. Some of the passages are bit too New Agey for me, but the core idea, expressed in the title, is clear and useful. A finite game is like a single game of baseball, with an end. An infinite game is like the overall phenomenon of baseball, which has no end. It is always a frontier.

So many utopian ideas are about finite games: End disease, solve global warming, get people to be more rational, reduce violence, and so on. As wonderful as all those achievements would (will!) be, there is something missing from them. Finite-game optimism suggests a circumscribed utopia without frontier or mystery. The result isn't sufficiently inspiring for me, and apparently it doesn't quite grab the imaginations of a lot of other people who are endlessly fascinated by dubious religious and political utopias. The problem is heightened at the moment because there's a trope floating around in the sciences, probably unfounded, that we have already seen the outline of all the science we'll ever know and we're just in the process of filling in the details.

The most valuable optimisms are the infinite games, and imagining that new innovations as profound as language will come about in the future of human interaction is an example of one.

Universal Telepathy

RUDY RUCKER

Mathematician, computer scientist; cyberpunk pioneer; novelist; author
of *Mathematicians in Love*.

There will be an amazing new discovery in physics on a par
with the discovery of radio waves or the discovery of nuclear
reactions. This new discovery will involve a fuller understanding
of the level of reality that lies 'below' the haze of quantum
mechanics – suppose we call this new level the domain of the
subdimensions.

Endless free energy will flow from the subdimensions. By
using subdimensional shortcuts – akin to what is now called
quantum entanglement – we'll be able to send information over
great distances with no energy cost. In effect, the whole world
can be linked like a wireless network, simply by tapping into the
subdimensional channel.

This universal telepathy will not be limited to humans; it will
extend to animals, plants, and even insentient objects. Via the
subdimensions, you'll be able to see every object in the world.
Conversely, every object in the world will be, in some limited
sense, conscious, in that it will be aware of all the other objects
in the world. A useful corollary is that any piece of brute matter
will be a computer just as it is. Once we can reach into the inner
self of an object, we'll be able to program the behavior of things
like rocks or logs without having to attach any kind of micro-
processor as an intermediary.

Humans will communicate at a vastly enhanced level. At
present, I communicate an idea by broadcasting a string of

words that serves as a program for reconstructing one of my thoughts. Once we enjoy subdimensional telepathy, I can simply send you a link to the location of some particular idea in my head.

Machines will fade away – and, in particular, digital computers will be no more. The emerging interactions of Earth's telepathically communicating beings will become a real and knowable Gaian mind. And then we will become aware of the other, higher minds in our cosmos.

The Best Is Yet to Come

NICHOLAS HUMPHREY

School Professor at the Centre for Philosophy of Natural and Social Science, London School of Economics; author of *Seeing Red: A Study in Consciousness*.

If I had lived in the year 1007 and had been asked what I looked forward to for my descendants in the next millennium, I might have imagined many wonderful possibilities. But I would not – because I could not – have imagined the music of Mozart, the painting of Rothko, the sonnets of Shakespeare, the novels of Dostoyevsky. I would have failed to see one of the best reasons of all for being optimistic, which is the power of human artistic genius to astonish us again and again. I will not make the same mistake twice.

So let me say straight out: In 2007, I hope and expect that

the best is yet to come, that greater works of art than any the world has ever seen will be created by human beings not far ahead of us – works of currently unimaginable aesthetic and moral force. And, mind you, it will not require genetic modification, computer hybridization, high-tech brain enhancement or whatever – it will simply require that we continue to be the kind of people we are.

The Restoration of Innocence

ELIZABETH F. LOFTUS

Professor of psychology, University of California, Irvine.

'I don't think a lot of people realize how important innocence is to innocent people.' These are haunting words spoken in the acclaimed 1988 Australian film about wrongful conviction, *A Cry in the Dark*.

The wrongful conviction of innocent people has been a serious problem in our society, a problem we're now becoming acutely aware of through the release of convicted individuals who were shown by DNA testing to be innocent. One happy consequence of these sad cases is the advent of a number of 'innocence projects,' typically operated out of law schools and dedicated to the freeing of those wrongfully convicted.

I wish I could say I was optimistic that wrongful convictions will virtually disappear, sort of like polio. I can't. But I *am* optimistic that the problem of wrongful conviction will be ameliorated. Here's why. Just as a plane crash leads to micro-analysis of what went wrong, so these cases of proved wrongful conviction have been dissected to determine what caused them. The answer, in most cases, is faulty memory. Mistaken identification by rape victims and others comes as no surprise to scientists who have studied eye-witness memory. We have learned a great deal about what it is about our system that promotes these tragic errors. (Readers can find out how these kinds of errors happen by reviewing the cases on the Web site of the Innocence Project.)

And finally our government is listening – the price paid by hundreds of individuals who suffered through years of imprisonment and are now free. The Department of Justice has convened a committee to make recommendations to law enforcement for how witnesses and victims should be handled to preserve valuable 'memory evidence.' Many states have recently adopted a package of reforms for how witnesses are interviewed and lineups are conducted. It has been a triumph of scientific discovery – a science that has taught us much about the workings of the human mind – and has also made a difference in the way our world works. But science has only scratched the surface and has layers upon layers to go. We will begin to see more memory science, more reforms in the justice system, and fewer errors. As we invest in the science, we need to keep one important idea in mind: Memory, like liberty, must be cherished, nourished, and protected. Without one, we can easily lose the other.

I Will Be Dead Wrong Again

THOMAS METZINGER

Philosopher, Johannes Gutenberg-Universität, Mainz; author of *Being No One*.

I am optimistic that I will be dead wrong again. As I have frequently been in the past. Being a philosopher, I was strongly opposed to marriage – on strictly theoretical grounds, of course! And about the only thing I always agreed on with Nietzsche was that married philosophers basically are clowns, as he put it – people who belong in a comedy. Real life proved me wrong (at least *I* think so), and I am glad it did.

Not a single one of our high-paid sociologists and politicologists predicted that the Berlin Wall would come down in 1989. They were dead wrong. And, boy, would each one of them have *loved* to be the one to make exactly this prediction!

I was also dead wrong in believing that European governments would never have the guts or the power to ban advertisements for tobacco products – or that European citizens would actually stop smoking in bars and public places simply because their governments told them to. Wasn't it much more plausible to expect major rebellions in countries like Ireland or Italy? How could anyone believe this would actually work?

Now that America is not a Western country anymore, I have serious doubts that Europe can actually rescue the heritage of the Enlightenment. Who will sustain democratic values and fight for all those old-fashioned ideas, like human rights and freedom of speech? China looks for a path of its own, but in a way that many find unsettling. Will India – now the world's greatest democratic project – manage, or will it collapse into even more corruption and self-generated chaos? Who will conserve and cultivate our legal systems, who will culture scientific rationality and the brand new tradition of settling ethical issues by argument rather than by force?

Europe is in a strange condition: Russia looks bad, Italy is a twilight state, Germany can't move; in many countries, such as Austria and Denmark, the voters are flirting with the extreme right. No Constitution. No common vision. And the pressures of globalization on our social nets.

If global sociodynamics keeps on unfolding as it currently does, isn't it likely that Europeans, on their own, will be powerless to alter the overall trend? America is gone for good. How rational is it, really, to keep believing that Europe will not only accept the historic challenge but eventually get its act together?

I am optimistic that, once again, I will be dead wrong.

The Modeling of Group Behavior

PAMELA McCORDUCK

Author of *Machines Who Think*.

At seventeen I saw that contemporary literature – I studied it then, and hoped eventually to be part of it – is an abyss of despair. No surprise: It reflects the unspeakable circumstances of the 20th century. Even so, it's no good thing to be seventeen and without hope. Luckily, chance brought me together with some scientists, and I discovered that in science, optimism was, and is, abundant.

Since then, I've spent much of my life trying to persuade my friends in the humanities that optimism on behalf of the human condition is a plausible point of view. It isn't the only point of view – the past century's horrors are a fact, and most of them happened through human agency. But the full life can support several points of view, often simultaneously, and my personal inclination is toward optimism, however qualified it must be.

For a long time my optimism centered on computing and what kinds of benefits it might bring us. Events have shown that I entertained far too modest an optimism. I'm embarrassed to say that the impact of the Internet – in particular, the World Wide Web – eluded me completely at first. A few years ago, I returned to artificial intelligence, which I'd written about early and then gone away from. Press narratives were uncritical about the field's death throes, and I expected to write an elegy. Instead, I found a revelation. Artificial intelligence is not only robustly healthy, building on its very significant gains since I first wrote about it, but the field's present ambitions burst with, well, *vitality*.

Lately I've been examining a new aspect of computing, the modeling of human behavior in groups, small and large, beginning from the bottom up, playing out dynamically, as only computer models allow. Years ago, in a casual dinner conversation with a social scientist, I wondered aloud if what prevented us from understanding what really happens to humans in groups is that we haven't found the code. I meant to make a vague comparison between the genetic code and something hypothetical that encoded human behavior. Instead of laughing, she solemnly agreed.

Such a code is not yet on the horizon, but thanks to some

marvelous new work by gifted social scientists, its intimations are teasing us. I'm optimistic that it will eventually be found. When it is, it will be a scientific triumph. It will open not just the future to our understanding but also the past. It will be a human triumph.

What will it mean to have such a code? For one thing, we can plan more intelligently. Want to wage a war? Call in the experts to run a few scenarios for you, laid out in bottom-up detail, humans and their interactions with each other and the terrain they're going to fight it out on. Watch the silicon agents melt away, to fight you another day; watch them reach out for help elsewhere. Once you see the model run its course, you may not want to fight that particular war after all.

Want to predict the possible spread of a disease? Good, the silicon model will tell you how many will fall, and where you can intervene to pinch off contagion effectively, and where it's a waste of effort. Want to figure out the ebb and flow of urban crime waves and then how to prevent them? Play it out in silicon first. Why do humans cooperate at least as much as they compete? Compare identical silicon societies – same people, same resources – but vary the amount of cooperation, the amount of competition. Which society collapses? Which society survives? Which society thrives? Where's the tipping point?

Perhaps as interesting, we'll be able to reach backward in time. How, really, did Mesopotamia become a desert, when once it had supported a network of rich societies? How much of that collapse was climate change, how much human folly? Build a model of early modern Europe and show what really caused the European Renaissance. Compute in detail how Great Britain came to rule the waves – and then didn't anymore.

We assume we've solved some of these problems, though

historians dispute one another ferociously, as do epidemiologists, as do economists – sometimes over details, sometimes over emphasis, sometimes over fundamental assumptions. Here comes a chance to nail it down, and these techniques offer us insights we couldn't get any other way. Finding the code that I once thought was only hypothetical will revolutionize our view of who we are, how we got that way, and who we might become, just as cracking the genetic code revolutionized biology.

And I'm chronically too modest in my hopes, so you can comfortably hope for more.

Assistive Listening

DAVID G. MYERS

Social psychologist, Hope College; author of *A Quiet World: Living with Hearing Loss.*

I foresee a friendlier future for us 31 million Americans with hearing loss. It's no news that cochlear implants, which were unavailable to my mother, should spare me the deafness that overtook her late in life. But few people are aware that many more of us could benefit by doubling the functionality of our hearing aids.

We can dream of a future where hearing aids serve not only as sophisticated microphone amplifiers but also as wireless loudspeakers that deliver clear, customized sound. In theaters, auditoriums, houses of worship, airport lounges, drive-up order stations, and home TV rooms, sound signals will be transmitted via in-the-ear loudspeakers, much like wi-fi transmissions to laptops.

Good news! That future has arrived in the United Kingdom and much of Scandinavia, and now in more than a hundred western Michigan facilities, and it is coming to several other American cities (www.hearingloop.org). When people experience public address or TV sound via 'hearing-aid-compatible assistive listening' (made compatible with the flick of a hearing-aid switch), they typically respond with amazed joy. What's more, they report increased satisfaction with their hearing aids.

It's a challenge to persuade a nation to exchange its current hearing-assistive technology (which requires locating, checking out, and wearing conspicuous headsets) for a technology that many more people would actually use. But the results of our western Michigan experiment, and another in 1,000 California homes, supports my optimism. Doubling hearing-aid functionality will greatly increase hearing-aid acceptance and use. With on-the-horizon technology, we can also foresee music buffs with wireless ear-bud loudspeakers. When that day comes, having something in one's ear will become as mundane as glasses for the eyes, and millions of people with hearing loss will be enjoying fuller and more connected lives.

We Will Find New Ways to Block Pessimism

RANDOLPH M. NESSE

Professor of psychiatry and psychology, University of Michigan; coauthor
(with George C. Williams) of *Why We Get Sick: The New Science of
Darwinian Medicine*.

I am optimistic that we will soon find effective new methods for
blocking pessimism. We are well on the way with antidepres-
sants. Side effects remain a major problem, and some people do
not respond, but progress has been rapid. Findings from neuroscience and genetics will provide the foundation, but the
engine that will drive new developments is the huge profit
potential from agents that relieve negative emotions. The anx-
iety- and depression-drug market already tops $20 billion per
year just in the United States.

I am pessimistic, however, about our ability to use those
drugs wisely. Pessimism is not a problem, it is a useful emotional state. When the boat overturns a mile out to sea,
optimism about one's ability to swim to shore is deadly. When
a hurricane is approaching, optimism is fine nine times out of
ten, then comes Katrina. When deciding whether to invade a
foreign country, optimism about receiving a warm welcome
can result in a catastrophe that changes the whole course of history for the worse.

The tendency to view optimism as superior to pessimism is a
deep-rooted illusion. Optimism is useful in propitious situations; pessimism is useful in dangerous situations. For the
fortunate, life now is vastly safer and more secure than it once
was, so pessimism is less necessary. But unintended consequences

of blocking pessimism are likely. Already, thousands of employees are subjected to motivational exercises to foster positive thinking. What will happen when we can all choose to feel positive most of the time? The world will be better in many ways and worse in others that are hard to predict.

The Limits of Democracy

MARK PAGEL

Evolutionary biologist, Reading University, U.K.

Some historians think the idea of democracy arose among the Greek warriors of the seventh to fourth centuries BC who manned the trireme warships. Up to sixty men – a deme – rowed these daunting three-tiered ships. Their effectiveness in battle depended upon precise and coordinated teamwork; the phrase 'pulling together' may have its origin in the triremes. Demeocracy arose when the rower-fighters realized that the same kind of coordinated pulling together that powered the boat could be used to influence which battles their masters had them fight and the conditions of their service. Herodotus records that up to forty triremes were used when Samos invaded Egypt – a lot of voting oarsmen. Modern democracies owe a debt to the actions of these wretched fellows, whose fates were gambled by rulers who did not always have the rowers' best interests at heart.

In spite of this, and 2,500 years on, I am optimistic that the world is glimpsing the limits of democracy. I speak of democracy in its wider manifestations and not just as government. The common idea of democracy – that everyone has a 'right' to be heard – naturally flourishes among the smallest collections of people who can organize themselves into a group. To survive, these groups compete for what they see as their share of the pie. Look no further than the ultra-democracies of some Western European nations, deadlocked coalition governments legitimized by systems of proportional representation that reward small, special-interest demes. Look to the intolerance that arises when this or that group asserts rights over other groups' rights. Look to 'focus groups.' Look to a state of numerical populism in which the most votes or text messages or viewers is what is to be delivered, crowned, or sold. In the artist Paul Klee's words, 'Democracy with its semi-civilization sincerely cherishes junk.' Strong words perhaps, but there is that old saying about decision-making by committee.

The British playwright Dennis Potter, in a public speech not long before his death, defended the BBC – a decidedly undemocratic and state-owned institution – for its very lack of populism. To Potter, the role of the BBC was to decide for the rest of us the standards of what should be deemed good art, drama, history, and reporting, and to challenge us intellectually and aesthetically. A dangerous state of affairs? Elitism? Maybe. But I am optimistic that people are recognizing that democracy simplistically applied can easily descend into an ochlocracy – government by the mob – ruled by the groups that shout the loudest. As early as the mid-19th century, we have Disraeli saying, 'The world is wearied of statesmen whom democracy has degraded into politicians.'

A hopeful sign: Some nations, and especially some American states, are researching new, low-carbon-footprint technologies and voluntarily committing to expensive climate-change targets.

In most cases, they are doing so without any democratic mandate. They realize that there may be larger and longer-term stakes to play for than the 'right' to behave as one pleases or to have what one wants. Maybe you know of other examples.

The World Is a *Wunderkammer*

DAVID PESCOVITZ

Coeditor of *BoingBoing*; research affiliate at the Institute for the Future; editor-at-large for *MAKE*.

Several years ago, I became fascinated with cabinets of curiosity. The Renaissance predecessor of modern-day museums, these cabinets, sometimes entire rooms, were filled with a mish-mash of objects both natural and artificial that embodied the wonder of the world. The German term for these collections is *Wunderkammer* ('chamber of wonders'). Inside, you might find a mummy's hand, a supposed unicorn's horn, exotic seashells from distant lands, odd insects pinned and catalogued, possibly even a two-headed lizard in a jar of formaldehyde. As Tradescant the Elder, one of the most notable cabinet keepers in history, requested in a letter to the secretary of the English Navy in 1625, this was a quest for 'any thing that is strang.'

Inspired by this celebration of science, art, and the strang(e), I picked up an old Chinese tea cabinet at a flea market and began to build my own *Wunderkammer*. I quickly filled the shelves with items of the type I thought should be in any

Wunderkammer worth its weight in weirdness: antique medical instruments, a primitive eye-gouging weapon from Rarotonga, a Balinese shadow puppet, a stuffed snake in a perpetual strike. Things got more interesting, though, once the collection process became more organic and I added items that genuinely spoke to my personal sense of the curious: a 1/1-millionth scale model of Frank Lloyd Wright's Fallingwater fabricated by engineers Ken Goldberg and Karl Bohringer using techniques borrowed from microscale manufacturing; a vial of carbon nanotubes; a Houdini automaton's autograph; a resin model of a telerobotic insect outfitted with solar cells for wings.

Now this small cabinet in the corner of my office serves as a constant reminder for me that the world is filled with wonder, and that curiosity is to be cultivated at every opportunity. Indeed, we're at our best when we're curious. And the beauty of curiosity is that we're all naturals. Curiosity is how babies learn. In fact, sparking someone's curiosity, at any age, seems to be perfect pedagogy. As the professor says in *The Day the Earth Stood Still*, 'It isn't faith that makes good science, it's curiosity.'

I wouldn't dare suggest that there's a Renaissance revival afoot, but I'm optimistic that the pendulum is swinging at least slightly back toward the heyday of natural history, citizen science, backyard astronomy, and other spirited intellectual pursuits. Several recent museum exhibitions have explored the cabinets of curiosity as an organizational principle, including one dedicated to the appropriately odd juxtaposition of art and cryptozoology. The *Wunderkammer* aesthetic has bubbled up into popular consciousness.

Many blogs, including the one I coedit, have been described as virtual cabinets of curiosity – storehouses of unusual links, odd memes, fringe culture, and weird news. Nearly every major city has at least one carefully curated Olde Curiosity Shoppe selling strange *objets d'art* and natural oddities packaged as Victorian chic. In fact, I was recently struck by the obviously *Wunderkammer*-inspired display of mounted insects and red coral on sale at a mainstream home-decor store in the mall. And, as ultimate evidence of a trend, at least two coffee-table books on the subject have been published in the last few years.

Most of all, I'm heartened by the unbridled curiosity fueling today's passionate do-it-yourself movement. A growing number of ingenious individuals are hacking Priuses to boost the gas mileage, installing Linux on iPods to record high-quality audio, and building backyard weather balloons. These makers are dissatisfied with off-the-shelf products. At a deeper level, they're driven by a daring inquisitiveness about what lies under the hood of today's technology and how they can improve what they buy or build it from scratch. For these makers – in the tradition of crafters, tinkers, scientists, engineers, artisans, and hot rodders who came before – the process is the product. The fun is in the fix. No user-serviceable parts inside? Says who?

I'm optimistic that in the coming few years the DIY movement will achieve not only widespread awareness but widespread participation. I'm optimistic that smart companies, instead of criminalizing hackers, will encourage these user-innovators and solicit their feedback to design better products. I'm optimistic that science education in the United States can be saved if students are given the opportunity to learn by doing, not just by reading about what someone else has done.

When I watch a screwdriver-wielding maker eagerly voiding another warranty, I see a spark of the same childlike curiosity that fills a baby's eyes as he or she first explores the world, optimistic that something wonderful lies ahead.

Overcoming the Burden of Monocausalitis

ERNST PÖPPEL

Neuroscientist, chairman of the board of the Human Science Center and chair of the Institute of Medical Psychology, Ludwig-Maximilians-Universität, Munich; author of *Mindworks: Time and Conscious Experience.*

Can I, as a scientist or a citizen, be optimistic about such questions as: Can we come to sustainable peace? Will we really solve one day the question of how our brains function? Are we going to win the battle against diseases? Will we ever be free of prejudices? And so on. The answer is an emphatic 'No.' There is no reason to be optimistic about such big questions.

On the other hand, I think of myself as an optimistic person. On a personal level, I am optimistic about the future of my children and grandchildren, about the career of my past and current doctoral students, about the realization of new research projects in the near future, about my health, and so on. If everybody were able to be this optimistic about personal concerns, there could be reason for an optimistic attitude toward the big picture. Such optimism would be an expression of trust, not of confidence that we can solve the problems of humankind.

I wish I could be optimistic about conquering a disease common to us all – namely, monocausalitis. Humans tend to want to explain everything in a monocausal way; we are always looking for one reason for things. The philosophical sentence *Nihil est sine ratione* ('Nothing is without reason') is usually misunderstood as 'Nothing is without one reason.' Ockham's razor – the choice of the simplest solution to a problem – is OK as long as the solution is not *too* simple.

In understanding biological processes – for instance, brain processes and how they control the 'mindworks' – we should free ourselves from this monocausal trap. I refer not only to the problem of the many hidden variables in analyses of biological processes – which, since they cannot all be controlled, are a headache for an experimenter – but also to a structural problem. Biological phenomena can be better understood if multicausality is accepted as a guiding principle. In particular, I would like to promote complementarity as a generative principle. In quantum mechanics, to the best of my knowledge, complementarity is a descriptive principle; in biology it is a creative principle. Just one example: It doesn't make much sense to explain human behavior on a genetic basis alone; genetic and environmental information together form (for instance) the matrix of our brain. This and many other examples are so self-evident that it is hardly necessary to mention them.

But if one looks at the optimism expressed about saving our world or achieving ultimate insight into Mother Nature's tricks, one is confronted with monocausal solutions. If we recognize in our evolutionary heritage the burden of monocausalitis, we may overcome this disease, at least partly.

Things Could Always Be Worse

ROBERT R. PROVINE

Professor of psychology and assistant director of the Neuroscience
Program at the University of Maryland Baltimore County; author of
Laughter: A Scientific Investigation.

Things could always be worse. Is this a cause for optimism? And
if so, is this a form of optimism worth having – a wimpy, agnos-
tic, noncommittal, damn-with-faint-praise kind of optimism?
Quite the contrary. It's a rough-and-ready, robust kind of optimism. This optimism is suited for everyday life and doesn't fold
under pressure. We have all heard of the 'grass is greener' syndrome. This is its 'grass is browner' counterpart, the achievable
antidote for broken dreams and bolsterer of the status quo.

Psychophysics – the study of the psychological effects of
physical events – indicates that more is not always better and
that greener grass, once acquired, quickly starts to yellow. While
two inches always seem twice as long as one inch, other sensa-
tions do not grow in a linear manner. A tone, for example,
must be much more than twice as powerful to sound twice as
loud to us. As with tones, the quirks of our brain doom a path
to happiness based on the accumulation of stuff. The second
million dollars, like the second Ferrari, does not equal the satisfaction provided by the first, and a second Nobel is pretty
much out of the question, a dilemma of past laureates. Goals
once obtained become the new standard, to which we adapt
before continuing our race up the escalating slippery slope of
acquisitiveness and fame. Philosophers and scientists from
antiquity to the present generally agree that life is a marathon,

not a sprint, and the formula for happiness and well-being is the journey, not achievement of the goal – and the comfort of friends and family.

This brings me back to my proposal: 'Things could always be worse.' It finesses our biologically determined law of diminishing returns and the impossibility of keeping up with the Joneses. Lacking the understated nobility of 'We have nothing to fear but fear itself,' my slogan would not lift the spirits of a Depression-era nation, serve a candidate seeking political office, or provide a philosophy of life, but it does help me to slog on. Best of all, my modest proposal is unconditionally true for anyone healthy enough to understand it; when things take a nasty turn, as they often do, celebrate the present and repeat it to yourself.

The Future

MATT RIDLEY

Science writer; founding chairman of the International Centre for Life; author of *Francis Crick: Discoverer of the Genetic Code.*

The future. That's what I'm optimistic about. The historian Macaulay said in 1830, 'We cannot absolutely prove that those are in error who tell us that society has reached a turning point, that we have seen our best days. But so said all who came before us and with just as much apparent reason.' The eternal, enduring pessimism of human beings about the future does real harm by persuading people, especially the young, to retreat from adventure and enterprise into anomie. Sure, the world has problems: AIDS, Islamo-fascism, carbon dioxide. But I bet we can solve them as we have solved others, such as smallpox, the population explosion, and the high price of whale oil.

Humankind Is Particularly Good at Muddling

PAUL SAFFO

Technology forecaster; consulting associate professor, Stanford University School of Engineering.

I am a short-term pessimist but a long-term optimist. History is on my side, because the cause of today's fashionable pessimism lies much deeper than the unpleasant surprises of the last half-decade. In fact, both our pessimism and the rise of the fundamentalisms that so bedevil global society at the moment share a common source – the year 2000 and the rollover into this century. The approach of each new year inevitably, predictably, causes us to look back and wonder what lies ahead. Moreover, contemplation of the new year amplifies our predispositions; pessimists become more certain that things are falling apart, while optimists see greater hope than ever. Opinion inevitably clusters at the extremes. This tendency is amplified by the number of zeros in the year to come. The end of a decade affects the zeitgeist for a year or two, while century endings reverberate for ten years or more, as demonstrated by the impact of the *fin de siècle* a hundred years ago.

We have less experience with millennium rollovers, but we're learning fast. With perfect hindsight, the influence of the approaching millennium can be seen throughout the 1990s and even earlier. Millennial anxieties contributed in no small part to the rise of religious fundamentalism, while millennially inflated hopes encouraged the touchingly innocent optimism overlaid atop the Internet revolution and emergent globalization.

Inevitably, the greatest impact of our calendric angst occurs after the triggering date has passed. The year 2000 still affects – perhaps even dominates – the zeitgeist today. Eschatologically obsessed believers like Muqtada al-Sadr stand astride events in Iraq, convinced by the calendar that the Mahdi redeemer will finally reveal himself. Closer to home, an astonishingly large population of Americans are equally convinced that the Apocalypse will arrive at any moment, and there is little doubt that fundamentalist apocalyptic beliefs directly affect U.S. policy. There also is no shortage of millennially inspired optimists (some whose answers are found in these pages), confident that the wonder machines of science and technology will allow us to live forever, power our devices with puffs of hydrogen, banish terrorism, and usher in a new age of human understanding and world peace.

I am a short-term pessimist because the Millennium is still clouding our collective thinking and may yet inspire the addled few to try something truly stupid, like an act of mega-terror or a nuclear exchange between nations. But I am a long-term optimist because the influence of the Millennium is already beginning to fade. We will return to our moderate senses as the current uncertainties settle into a comprehensible new order. I am an unshakable optimist, because, in its broadest strokes, the future will be what the future has always been – a mix of challenges, marvels, and endless surprise. We will do what we have always done and muddle our collective way through. Humankind is particularly good at muddling, and that is what makes me most optimistic of all.

The Increasing Coalescence of Scientific Disciplines

GERALD HOLTON

Mallinckrodt Research Professor of Physics and Research Professor of History of Science, Harvard University; author of *Thematic Origins of Scientific Thought*.

Under our very eyes, research in science has been taking a courageous and promising turn, to realize in our time an ancient dream.

Since Thales and other philosophers on the islands in the Ionian Sea, over 2,500 years ago, there has been an undying hope that under all the diverse and fluctuating phenomena there could be found in nature a grand, majestic order. This fascination, the 'Ionian enchantment,' has persisted ever since, in various forms.

Thus Isaac Newton thought the mechanical forces that explained the motions of the solar system would turn out to run all else, including human senses. Many attempts were made to extend Darwinian natural selection to include all societal phenomena. The influential Austrian polymath Ernst Mach, to whom young Einstein referred as one of his most important influences, taught that the true task of scientific research is to establish a form of fundamental science, an *Einheitswissenschaft*, on which is based every specialty. From about 1910 on, an increasing number of scientists in Europe and America gave allegiance to the idea of the 'unity of science,' a widespread movement hoping to find functioning bridges not only between the sciences but also between science and philosophy – Niels Bohr being one of the prominent promoters.

By and by, it became clear that such hopes were at best premature, that there was not enough of what William James called 'cash value,' in terms of actual accomplishments – not least in attaining a unified field theory. At one of the last meetings devoted to discussions about the unity of science, in 1956, J. Robert Oppenheimer, with typical eloquence, offered a valedictory to the Ionian enchantment with these words:

> It may be a question [whether there] is one way of bringing a wider unity in our time. That unity, I think, can only be based on a rather different kind of structure than the one most of us have in mind . . . The unity we can seek lies really in two things. One is that the knowledge that comes to us in such terrifyingly inhumanly rapid rate has some order in it . . . The second is simply this: We can have each other to dinner. We ourselves, and with each other by our converse, can create, not an architecture of global scope, but an immense, intricate network of intimacy, illumination, and understanding.

But even as such opinions were accepted with resignation, something new had been born, quietly grew, and in our time has become the source of increasing optimism about the value of the old dream – by turning in a new direction. Scientific research (at first only sporadically during the last century but more and more in our time) has been successfully reaching out for a new sort of unity – *in practice*, for an integration among disciplinary fragments. This time, the movement is not driven by a philosophy of science or a search for the Ur-science. Rather, it is appearing as if spontaneously in the pursuit and progress of research science itself.

There is an increasing coalescence of scientific disciplines in many areas. Thus the discovery of the structure of the genome not only required contributions from parts of biology, physics, chemistry, mathematics, and information technology but also led to further advances in biology, physics, chemistry, technology, medicine, ecology, and even ethics. And all this scientific advance is leading, as it should, to the hopeful betterment of the human condition (as had been also one of the platform promises of the unity-of-science movement, especially in its branch in the Vienna Circle).

Similar developments happen in the physical sciences – a coalescence of particle physics and large-scale astronomy, of physics and biology, and so forth. It is a telling and not merely parochial indicator that about half of my forty-five colleagues in Harvard's Physics Department, owing to their widespread research interests, now have joint appointments with other departments at the university: with Molecular and Cellular Biology, with Mathematics, with Chemistry, with Applied Sciences and Engineering, with History of Science. A new building is being erected next to the Physics Department, called the Laboratory of Integrated Science and Engineering. Although in industry, here and there, equivalent labs have existed for years, the most fervent follower of the unity-of-science movement would not have hoped then for such an indicator of the promise of interdisciplinarity. But as the new saying goes, most of the easy problems have been solved, and the hard ones need to be tackled by a consortium of competences.

In other parts of this university, plans are under way to set up a program for higher degrees in the new field of systems biology, which has the goal of reaching an 'integrated understanding' of biological/medical processes. It will bring together faculty and students from biology, medicine, chemistry, physics, mathematics, computation, and engineering. These parochial examples indicate a general trend in many universities. The new password to success is 'integration' and 'interdisciplinarity.' If an official sacralization of this movement were needed, it would be the 2005 release of a big volume by the National Academy of Sciences titled *Facilitating Interdisciplinary Research*.

All this is not precisely what the philosophers and scientists, from Thales on, were hoping for. We will not – at least, not for a long time – have that grand coalescence of all sciences and more. These developments lack exalted philosophical pretensions; instead they are a turn to many-heads-together, hands-on work on specific hard problems of intense scientific interest, many of which are also of value to society at large.

And, of course, these co-workers can also still have one another to dinner.

The End of Hegemonies

BARRY C. SMITH

Philosopher, School of Advanced Study, University of London; coeditor of
Knowing Our Own Minds.

What I am optimistic about is that ultimately monopolies fail.
By which I mean attempts to dominate our tastes, our preferences, our culture, our media, our political policies, our moral
choices. Restless creatures that we are, we seek out variety and
difference, opportunities to extend the scope of our thinking
and exercise discrimination and taste. This may make us hard to
satisfy, but it is this lack of satisfaction that leads to progress and
spells the end of hegemonies in ideology, religion, or science.

John Stuart Mill wondered whether each of us would rather
be the pig satisfied or Socrates dissatisfied, and at times it may
seem as though a lot of people have chosen the former. But that
is only in the short term. Long-term, we have no choice but to be
dissatisfied when things are constant and unchanging. The sati-
ety of our appetites, the endless repetition of the same thoughts
and feelings, will eventually, in all but pathological cases, lead us
to move on in mind and seek fresh inputs. To begin with, people
may readily sacrifice their freedom for comfort, but increasingly
the absence of change, the monotony of surroundings and rou-
tines, will lead to acute discomfort and the search for something
new. That is why I am optimistic that people who are fed a constant diet of the same ideas, the same foods, the same TV
programs, the same religious or political dogmas, will eventually
come to consider other possibilities, will switch off, change allegiance, and think differently and for themselves. It may take

time; after all, some people's threshold for boredom is higher than others. But change and a moving on will be inevitable. The lesson is already being learned in the corporate world, where monopolies try to cope by diversifying their range of services. Their chance of survival will depend on how cynically or sincerely they respond to this restless aspect of the human mind. We are used to hearing how bad the diet of television or Hollywood movies is, and how people have come to expect less and less. But I think the opposite is true. People are increasingly switching off and staying away from the familiar and undemanding shows and films that lazy television executives and film producers offer. Instead, space has opened up for intelligent and entertaining programs and for independent filmmaking. It is here, at the creative end of the culture, that big popular success is to be found. In a similar vein, the increasingly global market has led to a firmer appreciation of the interesting local ones. And I am optimistic that people, through boredom and the need for something new, will seek out better, not worse experiences.

Human cognition depends on change and movement in order to function. Evolution has built us this way. Try staring at a blank wall for several seconds without blinking and you will find the image eventually bleaching until you can see nothing. The eye's visual workings respond to movement and change. So, too, do the other parts of our cognitive systems. Feed them the same inputs successively and they cease to produce very much worth having as output. Like the shark in water, we need to keep moving or, cognitively, we die.

Science, too, represents the greatest advert for our unquiet natures. For as soon as a theory or school becomes the established orthodoxy, creative minds begin to explore the possibility that we must begin from completely different starting assumptions and seek novel interpretations of the data. Without this constant movement to resist acceptance and stasis, we would not have the advances or excitements that fundamental science can provide. That said, we must not overlook the role that luck plays in great discoveries. But even with a lucky finding, we must be capable of recognizing and seizing on it if we are to develop insight for large-scale revisions to our thinking. The possibility of revising, reworking, and reconsidering depends on this sometimes uncomfortable fact about our natures.

So far, I have stressed the positive aspect of the restless mind, but there is a paradox in our nature and our restless search for change. For unless we countenance change for change's sake, or the relativist doctrine that anything goes (and I don't), how do we preserve the very best of our thinking, select better-quality experiences, and maintain our purposes and values? How do we avoid losing sight of older wisdom while rushing toward something new? It is here, perhaps, that our need for variation and discrimination serves us best. For the quick and gimmicky, the superficially appealing but weakest objects of our thinking or targets of desire will also be the least substantial and have an essential blandness that can tire us quickly. Besides, the more experience we have, the larger the background against which to compare and judge the worth or quality of what is newly encountered, and the more able we are to decide whether it will be ultimately rewarding. Certainly, people can be fickle or stubborn, but they are seldom fickle or stubborn for long. They will seek out better, according to what they are capable of responding to, and they will be dissatisfied by something not worthy of the attention they are capable of. For this reason, attempts to dictate tastes, cultural goods, ideologies, or ideas are bound in the end to fail, and about that, and despite the many dark forces around us, I am optimistic.

Understanding Sleep

STEVEN STROGATZ

Applied mathematician, Cornell University; author of Sync: How Order Emerges from Chaos in the Universe, Nature, and Daily Life.

In the coming year, we're going to witness a breakthrough in our understanding of what sleep is for.

It surprises most people to learn that this is even a question. Every other basic bodily function – like eating, drinking, breathing, urinating, or defecating – has a pretty clear purpose. Our bodies and brains need food, water, and oxygen to stay alive and replenish themselves, and if they didn't rid themselves of the by-products of this metabolism we'd be awash in our own toxic waste. Likewise, sleep must be for something important. We all spend decades in this strange state, immobilized, unconscious, and vulnerable. But what exactly does sleep do for us?

Children are told that they need to sleep because they're tired and need rest. But of course, rest is not good enough. Lying still for eight hours is no substitute for sleep. My mother had a different theory; she said I needed to sleep because I had too much 'sleepy gas.' It had been building up all day long, and I needed to sleep to get rid of it. In fact, scientists observed a long time ago that if you keep a sheep awake for several days and then inject some of its cerebrospinal fluid into another, well-rested sheep, that sheep will fall asleep immediately, presumably because some naturally occurring sleep substance had reached a soporific level in the donor. But this line of research never quite solved the puzzle. Although a number of putative sleep substances have now been identified, we're not sure how they might work biochemically, or how sleep (as opposed to mere rest) might break them down.

Other sleep-deprivation studies, done in the early 1980s, took a more brutal approach, keeping rats awake for weeks until they died from sleep deprivation and then looking for the precise cause of death. Such studies (now outlawed) could not pinpoint any specific culprits, such as particular organ failures. One striking observation, however, was that the rats ate much more than normal and yet wasted away. Their metabolism seemed to be wrecked. So maybe sleep is for energy regulation in some unspecified way. Other popular theories are that sleep is for tissue repair, or immune function, or for consolidating learning and memory.

The new development, and the cause for optimism, is an original approach to the problem, which makes the first quantitative, testable predictions about the function of sleep. Two physicists – Van Savage, of Harvard Medical School, and Geoff West, of the Santa Fe Institute – have analyzed how sleep varies across mammals of various species. Normally, physiological time ticks slower for bigger animals: For example, elephants live much longer than mice, and their hearts beat much slower. The interesting thing is that both animals' lifetimes and pulse times scale in the same way with their body mass – in direct proportion to their mass raised to the 1/4 power – with the curious implication that the hearts of mice and elephants will typically beat the same number of times in their lifetime.

What is so strange about sleep in this regard is that it behaves differently from all other physiological times. It's backward. Mice sleep longer, not shorter, than elephants do – about fourteen hours a day, compared with four for elephants. Savage and West interpret this as evidence that sleep is related to cellular repair. They note that cells produce destructive by-products, such as free radicals, in the course of normal metabolism, and they hypothesize that sleep somehow helps repair the damage that ensues. In this view, the mouse needs to sleep longer to clean up all the by-products generated by its revved-up metabolism. Then, using classic laws about how metabolic rate varies across different species, they derive mathematical predictions about how sleep duration should vary with an animal's size. But which size is most relevant – brain size or body size? The key is that they are not proportional. If sleep is for repairing the brain, Savage and West derive one prediction; if it's for repairing the body, they derive a different prediction. When they finally confront their model with the available data, they infer that sleep is mainly for repairing the brain, not the body. So much for beauty sleep.

Shortening Sleep Will Enrich Our Lives

MARCEL KINSBOURNE

Professor of psychology, The New School; coauthor (with Paula Kaplan) of *Children's Learning and Attention Problems*.

Our life span is extending, but the extended life is a dwindling asset. Who would not prefer to live longer while at his or her peak? The time we spend asleep contributes little to our lifetime of experience, thought, and action. Dreaming doesn't seem to add much. Some brain lesions and monoamine-oxidase (MAO) inhibitor medications even completely abolish dreams without making any apparent difference. Could we reduce the duration of sleep (both REM and non-REM) while maintaining its benefits for the brain, whatever they might be? I propose that we do need to sleep, but not as long as we do. The duration of sleep may be an outdated adaptation to prehistoric ecological constraints that no longer exist.

Virtually all vertebrates sleep (and invertebrates at least have quiet time). However, the duration of sleep varies wildly across species, from less than an hour to eighteen-plus hours a day. For instance, rodents sleep between eight and seventeen hours, primates between seven and eighteen hours. Elephants and giraffes sleep from three to five hours, squirrels for sixteen or seventeen hours, and bats twenty hours. The newborn of most species sleep more of each day than the adults, except that newborn whales and dolphins don't sleep at all. Within a species, the variation of adaptively valuable traits is thought to be quite limited – yet some people, in some families, habitually sleep only two to four hours a night and function well for longer each day. Perhaps constraining the duration of sleep is not an adaptive priority in humans.

Three categorically distinct roles for sleep are: (1) maintaining the neuronal circuitry, (2) fostering learning, (3) keeping the organism out of trouble.

(1) Given its ubiquity among vertebrates and other phyla, any neurometabolic benefit of sleep must be general and basic. If the needs of the brain determine the duration of sleep, its duration should vary systematically with some fundamental neurological variable, such as absolute or relative size of the brain, its energy utilization, the sophistication of behavioral

control, or the need to replenish some key neurotransmitter. No such covariation appears to exist.

(2) The presumed role of sleep in learning is based on continuing rehearsal. Rather than being an adaptation, the learning benefit may be a fortuitous result of the brain's continuing activity during sleep, while it is receiving no fresh information. Since the neuronal show must go on, recently acquired patterns of firing gain priority; whether the memories are useful or useless, they are automatically rehearsed. In any case, the suggested benefit of sleep for human learning cannot be generalized to species that make a living without learning anything much and yet require sleep.

(3) The substantial differences between people and the enormous difference between species in how long they typically sleep suggest that sleep also serves a species-specific ecological function. This is sleep's other role: Sleep conserves energy and keeps animals out of trouble. It takes the members of each species a minimum time per day to make a living – that is, to secure their personal survival and take advantage of any reproductive opportunity. This challenge is met anew every day. On this view, how much of the day is needed to meet adaptive goals determines the duration of the default option of sleep.

Continued activity when the day's housekeeping is done would prolong the animal's exposure to the hazards that lurk in the environment, without contributing further to basic survival and reproductive needs. Many species cannot do anything useful in the dark (and some not in the light). They gain nothing from expending more than basal metabolic energy at that time. The genetic imperative to sleep during a predetermined time of day and for a predetermined duration (or even hibernate) takes care of all that. Thus extended sleep time would be a function of the interaction between the individual and its ecology.

Predators need time for hunting; how much depends on attributes of the predator and the prey, such as speed, strength, and population density. Herbivore prey needs a minimum time to graze, depending on the animal's bulk and the accessibility of food. How the remains of the day are spent would depend on how readily a secure haven can be found.

Nature is notoriously conservative, and it conserves the genetically driven imperative to sleep. The imperative to sleep is subjectively experienced as antecedent sleepiness and the fatigue and dysphoric feeling after too little sleep. My thesis is that these feelings do not arise unavoidably from the economy of the brain but are genetically imposed adaptations. Should a species' ecology undergo radical change, and should making a living become remarkably easier or more difficult, natural selection will in time reshape sleep duration accordingly.

However, human culture evolves too quickly. Since artificial lighting was introduced, the dark no longer constrains what people can do. Because human activities are oriented as much to future as to immediate goals, all hours of the day have become potentially useful. Further, we have more effective means to secure ourselves than curling up in a quiet place and sleeping. So if a sizable portion of the adaptation to sleep has the role of a security-saving placeholder, then it would be safe to relax that portion of the sleep constraint.

The dictatorial 'sleep genes,' when identified, need to be modified to require a shorter sleep duration, and the circadian-clock genes need to be reset. Will the state of genetic engineering become sufficiently advanced to make this prospect, though less than a sure thing, more than a pipe dream? The good news comes with the fruit fly's sleep, which is uncannily like ours; a mutation in a gene called Shaker reduces the fly's natural sleep duration by two-thirds, from about twelve to about four hours within twenty-four, without detriment to the fly's well-being. The bad news is that these mutated flies don't live long. Nonetheless, I am optimistic.

The Joys of Failing Enterprises

MICHAEL WOLFF

Columnist, *Vanity Fair*; author of *Autumn of the Moguls*.

The good news where I come from, as a prisoner here in the American media business, is all about entropy. The massive and ridiculous systems we've built to control and market expression and culture, with their dissipations of so many people's energy, are deteriorating and coming apart. Every day, the media – a much more vexing monolith than religion and God – takes another transforming step from consolidation and uniformity toward uncertainty and chaos. This is not just because new technologies are revolutionizing production and distribution – though that's no small part of this creative destruction – but as much because of the inevitable disorder and randomness in closed systems. What's optimistic is that so many strategists and consultants and bureaucrats and moguls have not been able to maintain control and have been shown to be as clueless as everybody else. (This is true, come to think of it, not just at Time Warner and Viacom but at the Pentagon.) Breakdown can be a positive development.

Copying Is What Bits Are For

CORY DOCTOROW

Science fiction writer; blogger; technology activist; coeditor of *BoingBoing*.

I'm optimistic that the risks of anti-copying technology and the copyright wars are starting to move to the mainstream. Daily newspapers are reporting on the risks from Zune's DRM (digital rights management); governments and librarians are starting to question the fairy tales from the entertainment industry. The British government is poised to be the first in history to reject a proposal to extend copyright. A Canadian MP lost her seat last year because she'd sold out the country to a bunch of entertainment dinosaurs. Four European nations opened inquiries into the competition and consumer-protection issues raised by iTunes DRM. The latest WIPO (World Intellectual Property Organization) treaty looks moribund, killed by activist involvement.

Sure, the Free Trade Agreement between the United States and Russia restores the totalitarian practice of licensing presses

(Uncle Sam, bringer of liberty!) and plunges Russia back into the pre-*samizdat* era. Sure, the Recording Industry Association if America is continuing to terrorize American families by suing some 700 of them every month. But most people, confronted with the choice between HD-DVD DRM and Blu-ray DRM, are choosing none of the above.

There is no such thing as a copy-proof bit. There aren't even copy-resistant bits. Copying is what bits are for. They will never, ever get any harder to copy.

The copyright wars are a form of contemporary Lysenkoism, a farce wherein we all pretend that copy-proof bits are a reasonable thing to expect from technology. Stalin's Lysenkoism starved millions when the ideologically correct wheat failed to grow, and anyone who pointed this out was sent to prison. Entertainment industry Lysenkoism is ruining lives, undermining free speech and privacy and due process, destroying foreign democracies, and keeping poor countries poor. It's about time we wised up to it – and we are. That makes me optimistic.

Whether Solutions Are Possible

DAVID DEUTSCH

Quantum physicist, Oxford University; author of *The Fabric of Reality*.

They always are. Why is that important? First, because it is true. There is no anthropocentric spite built into the laws of physics, mandating that human improvement may proceed this far and no further. Nor is the dark, neo-religious fantasy true that nature abhors human hubris and exacts a hidden price that outweighs any apparent success, so that 'progress' always has to be in scare quotes. And second, because how we explain failure, both prospectively and retrospectively, is itself a major determinant of success. If we are optimistic that failure to improve ourselves means merely that we haven't found the solution yet, then success is never due to divine grace (nowadays known as 'natural resources') but always to human effort and creativity, and failure is opportunity.

Reforming Scientific and Medical Publishing via the Internet

BEATRICE GOLOMB

Associate professor of medicine, University of California, San Diego.

I am optimistic that the ascendance of open-access postings of articles on the Internet will transform scientific and medical publishing and that a number of profound problems – some particular to medical publishing – will be assuaged as a result.

Currently, it can be impossible to gauge the true balance of risks and benefits of medical treatments from a reading of the literature. Frighteningly, this is true too for those doctors who ground their clinical decisions on the literature. I will review some aspects of the problem and then relay grounds for possible optimism.

First, as is probably true in all fields, there is a bias in favor of existing orthodoxy. This is arguably more troubling in medicine, since the orthodoxy is itself influenced by the profusion of articles favorable to their products that are ghostwritten by the pharmaceutical industry or by the for-profit MECCs (medical education and communication companies) that industry hires for this purpose. These companies in turn pay physicians and pharmacists – including favorably disposed 'thought leaders' – to be the listed authors, extinguishing any appearance of connection to industry for the views propounded. This makes it seem as though independent parties are in agreement in their favorable representations of the evidence. Crisply put, advertising is published as though it were science.

These problems are worsened by direct conflict of interest, which is endemic in medical research. Articles about a class of drug are much more likely to be favorable when written by people with ties to industry than when written by those without such conflicts. Reviewers may also foster rejection of articles unfavorable to industry. (As elsewhere, reviewers are drawn from the pool of authors. Moreover, reviewers are seldom asked to disclose conflicts, and they remain anonymous, precluding disclosure of biased reviews.

These factors are aggravated, possibly dwarfed, by the influence of pharmaceutical companies on medical publishing, further aligning medical publishing with medical advertising. Medical journals are not the independent arbiters of article quality that one might wish; they are businesses and derive their revenue from pharmaceutical company advertising and from sales to industry of glossy reprints of industry-favorable articles, at inflated prices. For some medical journals, profits reportedly number in the millions, so the stakes are high.

At least three former editors-in-chief of major U.S. and British medical journals have written books decrying the inimical impact of industry influence on medicine. One has to ask why medical journals accept advertising (just because it is available for the taking) and whether the journal's bottom line is a proper consideration in dictating what is published, when lives are at stake.

So, whence the optimism? One way to propel optimism is to suggest a tactic that might justify it. Briefly, I suggest that papers be published on the Internet, reviews be submitted by named reviewers, and that others rate (and review) the reviews. Papers and reviewers alike will receive ratings, which can be continually updated. While this won't protect against biased submissions, it will protect against biased rejections – and at least enable a voice for original or contrary perspectives. It is probable that more bad science will be released; however, the system provides a means for improving poor quality work and avoiding what remains substandard.

More important, more good science may be published – perhaps more great science. As Nobelist Sydney Brenner (who famously wrote an article titled 'Moron Peer Review') has observed, much of his co-Nobelists' prize-winning work was initially rejected in the review process. Transformative work, by its nature, may defy conventional wisdom. One wonders: Is there other work that would have revolutionized science (and merited a Nobel) that languishes unpublished because authors ceased to persevere in submission efforts after some number of rejections?

Hark back to the many great discoveries of which we *have* heard that were initially ridiculed: *H. pylori* as a contributor to ulcers; hand washing as a means to reduce puerperal fever; the sun as the center around which the earth revolves, to name a few. What might this imply for the possibility that major discoveries may be pilloried into nonpublication by peer review? There is no way to estimate the fraction of Nobel-caliber efforts that achieve publication; the denominator remains unknowable.

The benefits of a new, Internet-based approach may be particularly great for the most important work: work that challenges existing orthodoxy; work that defines a new field and fits no existing journal; work that crosses discipline boundaries; science that is ahead of its time, that entails many advances at once or founds new work on an understanding of relevant material that others do not yet have. Or, too, work that runs counter to vested interest groups – particularly (but hardly exclusively) in the arena of medicine, where the potent impact of industry influence on information has been the subject of increasing alarm and where disparities between literature and truth may cost patients' lives.

An instance from mathematics supports the premise that requiring articles to be published in peer-reviewed journals may inhibit promulgation of at least some important work. The Poincaré conjecture – a holy grail in mathematics – was recently proved by a Russian mathematician who posted his work on the Internet but refused the bother of submitting his work to a journal. Other cases can be adduced favoring the proposition that some scientists capable of propelling major advances – which often entails rejecting convention – are constitutionally inclined to reject the petty obstacles and distractions that attend the current model of scientific publishing. And perhaps they do so with justifiable contempt.

Surely many will defend the current system – not least those who fare well within it and benefit disproportionately from it. And surely there will be problems to overcome in the new system. Orthodoxy, in-groups, and interest groups will continue to influence the literature. Those who serve these masters will likely submit negative reviews of articles (and of reviewers) who do not toe the respective party lines. But at least now the contrarian positions will achieve release, reviewers can be held accountable for biased reviews, and unacknowledged conflicts can be exposed, in instances when others know of them.

In short, I am optimistic that on-line publishing, with a review-the-reviewer system akin to that proposed here, will provide more voice and venue for science that may have the highest need – and the lowest prospect – of being aired.

The Real Purity of Pure Science

PIET HUT

Professor of astrophysics and interdisciplinary studies, Institute for Advanced Study, Princeton; coauthor (with Douglas Heggie) of *The Gravitational Million-Body Problem: A Multidisciplinary Approach to Star Cluster Dynamics.*

I grew up reading heroic stories about progress in science, the absolute superiority of the scientific method, the evil of superstition, and other one-dimensional optimistic views. Almost half a century later, I have a much more nuanced view of progress, method, and ways of looking at the world. What has been presented as *the* scientific method, at any given time, has been a simplified snapshot of an intrinsically much more opportunistic enterprise. As such, much damage has been done by suggesting that other areas, from social science and economy to politics, should adopt such a simple and continually outdated picture.

The strength of science is not at all its currently accepted method. The strength is the fact that scientists allow the method to change. The way the method changes is the exact same way that progress is made by applying the method in everyday research. Change of method takes place slowly and carefully, through long and detailed peer discussions, and may be almost imperceptible in any given field during a scientist's lifetime. The scientific method is like spacetime in general relativity: It provides the stage for matter to play on, but the play of matter in turn affects the stage.

The real basis for the success of science is its unique combination of progressive and conservative elements. A scientist gets brownie points for crazy new ideas, as long as they are interesting and stimulating – and also for criticizing new ideas, as long as the criticism appears valid. What is interesting in new ideas, and what is valid in the criticism thereof, is determined solely by peer review – that is, the collective opinions of the body of living scientists – not by falling back on some fixed notion of method.

Pure science has been held up as a beacon, a way to allow scientists to pursue their own intuitions and thus find new solutions to old problems. This is seen in contrast to applied science, where short-term goals do not allow sufficient room for finding really new approaches. The irony is that the best applications of science are ultimately based on pure rather than applied research.

The lesson has been that long-term research should not focus on goals but should let the scientific method follow its own course, free of goals. But I think this lesson is still misleading: The greatest breakthroughs have come from a doubly pure science, free of goals and method alike. Each major breakthrough was major exactly *because* it broke the rules – the rules of the scientific method as understood so far. The most spectacular example is quantum mechanics, which changed dramatically even the hallowed notion of experimental verification.

I am optimistic that all areas of human activity can be inspired by the example of science, which has thrived for more than four centuries without relying on goals or even on methods. The key ingredients are hypercritical but nondogmatic conservatism combined with wildly unconventional but well-motivated progressiveness. Insofar as there is any meta-method, it is to allow those ingredients to play off against each other in the enactment of scientific controversies until consensus is reached.

A Core Decency Even the Worst Government Machinations Can't Hold Down

DAVID BODANIS

Writer; futurist; author of *Passionate Minds*.

I'm optimistic because there's a core decency in people that even the worst machinations of governments can't entirely hold down.

The Evelina hospital is the first new children's hospital that's been built in London in a century. There's a giant atrium in the middle, and the contract with the company doing the cleaning says that the window cleaners need to dress up as superheroes. The children in bed – many with grave illnesses – delight in seeing Superman and Spiderman dangling just inches away from them, on the outside of the glass. Apparently for the cleaners it's one of the best parts of their week.

The government has wasted a fortune on consultants, bureaucracy, and reorganizations of the National Health Service. It's always defended in cold management-speak. This simple arrangement with the window cleaners cuts through all that. Everyone I've mentioned it to recognizes that – and in that recognition lies our hope.

The Rise of Autism and the Digital Age

SIMON BARON-COHEN

Psychologist, Autism Research Centre, Cambridge University; author of *The Essential Difference: Male and Female Brains and the Truth About Autism.*

Whichever country I travel to for conferences on the subject of autism, I hear the same story: It's on the increase.

In 1978, the rate of autism was 4 in 10,000 children; today (according to a 2006 article in *Lancet*) it is 1 percent. No one can explain this increase, though conservatively it is put down to better recognition and broadening the diagnostic category to include milder cases, such as Asperger syndrome. It is neither proved nor disproved that the increase might reflect other factors, such as genetic change or some environmental (e.g., hormonal) change. For scientists to answer the question of what is driving this increase will require imaginative research comparing historical and cross-cultural data.

Some may throw up their hands at this increase in autism. They may feel that the future is bleak for all of these newly diagnosed cases. But I remain optimistic that for a good proportion of them it has never been a better time to have autism.

Why? Because there is a remarkably good fit between the autistic mind and the digital age. The digital revolution brought us computers, but this age is remarkably recent. It was only in 1953 that IBM produced its first computer; a mere fifty-four years later, many children now have their own computer.

Computers operate on the basis of extreme precision, and so does the autistic mind. Computers deal in black-and-white binary code, and so does the autistic mind. Computers follow rules, and so does the autistic mind. Computers are systems, and the autistic mind is the ultimate systemizer. The autistic mind is interested only in data that are predictable and lawful. The inherently ambiguous and unpredictable world of people and emotions is a turn-off for someone with autism, but a rapid series of clicks of the mouse that leads to the same result every time that sequence is performed is reassuringly attractive. Many children with autism develop an intuitive understanding of computers in the same way that other children develop an intuitive understanding of people.

So, why am I optimistic? I anticipate that many among this

new generation of children with autism will find ways to blossom, using their skills in digital technology to find employment, to find friends, and in some cases to innovate. When I think back to the destiny of children with autism some fifty years ago, I imagine fewer opportunities for them. When I think of today's generation of children with autism, I do not despair. True, many of them will have a rocky time during their school years, when their peer group shuns them because they cannot socialize easily. But by adulthood, a good proportion of these individuals will not only have found a niche in the digital world but will exploit that niche in ways that may bring economic security, respect from their peers, and a sense of worth because of the contribution they are making.

Such opportunities may be open only to those individuals with autism who have language and otherwise normal intelligence, but this is no trivial subgroup. For those more severely affected – by language delay and learning difficulties – the digital age may offer less, though even for this subgroup I remain optimistic that new computer-based teaching methods will have an appeal that can penetrate the wall that separates autism from the social world. The autistic mind – at any level of IQ – latches onto those aspects of the environment that provide predictability, and it is through such channels that we can reach in to help.

A Second (and Better) Enlightenment

IRENE PEPPERBERG

Research associate in psychology, Harvard University; author of *The Alex Studies: Cognition and Communications Abilities of Grey Parrots.*

Like some other respondents, I'm not particularly optimistic at the moment. Human civilization, however, seems to proceed in cycles overall, and I believe we are due – even if not quickly enough for my tastes – for a new positive cycle. Every golden age – the flowering of reason and good – has been followed by a withering, a decay, a rotting, a descent into superstition, prejudice, greed (pick your own favorite ill). Somehow, though, the seeds of the next pinnacle begin their growth and ascent, seemingly finding nourishment in the detritus left by the past. A particular civilization may end, but new ones rise to take its place. I'm optimistic that the current nadir in which we find ourselves (a world mostly heedless of ongoing genocides, global warming, poverty, and so on), or toward which we see ourselves heading, will lead to a renaissance, a new enlightenment, a profound global shift in the worldview for the better.

Corrective Goggles for Our Conceptual Myopia

COREY S. POWELL

Executive editor of *Discover*; adjunct professor of science journalism, NYU; author of *God in the Equation: How Einstein Transformed Religion.*

Broadly speaking, I am optimistic that the world's current crises look terrifyingly large mainly because of our conceptual myopia. It is practically a truism to say that every era tends to regard its troubles as uniquely daunting, but I think that accelerating news cycles make the current generation particularly prone to this error of judgment. Making my best attempt to put on corrective goggles and take the longer view, I see a half-dozen areas where we are on the verge of major advances in our ability to expand our control over our environment and ourselves, in ways that will be largely or entirely beneficial.

- I am optimistic that technology will soon show practical ways to eradicate the twin problems of carbon emissions and fossil-fuel scarcity. In the nearer term, carbon dioxide will follow the path of CFCs, acid-rain-causing sulfur oxides, and most automobile tailpipe emissions. Naysayers warned that all of these would be difficult and economically disruptive to tackle; in every case, the naysayers were quickly proved wrong. Carbon sequestration is the most obvious technology for offsetting carbon emissions. Here's a firm prediction: If the world's leading economies set tough emissions standards for CO_2, or establish a serious carbon tax, industry will find astonishingly inexpensive ways to adapt within a few years.

Farther ahead, new energy sources will begin to make serious contributions to the world economy long before fossil fuels run out. My bet is still on fusion energy, despite its perfect five-decade record of never fulfilling any of its promises. I seriously doubt, though, that commercially viable fusion energy will look anything like the huge and hideously expensive magnetic-confinement test machines (like the international ITER project) now being built or planned. More likely it will take the shape of a pulsed plasma gun or a field reversed plasma triggering exotic nuclear reactions that send protons flying through a copper coil; then you have direct electricity conversion, with no boiler, no steam, no turbine, no dynamo.

- I am optimistic that we are on the verge of developing tools to program biological systems as effortlessly as we program digital ones. Synthetic biology, a field spearheaded by George Church, Drew Endy, and Jay Keasling, will be key to attaining this goal – and it is already in transition from theory to reality. Rather than snipping genes from one creature and clumsily inserting them into another, future biotechnicians will consult a master database of DNA sequences and specify the traits they want, whether to insert into an existing organism or to create a brand-new one, designed from the ground up. (A corollary is that these tools will finally allow effective stem-cell therapy, which leads to a related prediction: Thirty years from now, the current agonies over the ethics of stem-cell therapy will look as quaint as the hand-wringing over 'test-tube babies' in the 1970s.) Synthetic biology in its fully realized form could also be a dangerous weapon. A related part of my optimism is that – like electricity, radio, and all genetic research so far – it will prove far more useful for positive applications than negative ones.

- I am optimistic that young adults today will, on average, live to a hundred and twenty and remain healthy and vigorous until their final years. Researchers like Leonard Guarente, David Sinclair, and Cynthia Kenyon are zeroing in on the chemical and genetic basis of aging. Immortality is a long way off, but drugs and genetic therapies that hold back age-related diseases are coming soon. Treatments that slow the aging process as a whole will follow closely behind. Ultimately these will lead to a wholesale reordering of the pace of life and the familiar social structures based around certain biological milestones. The child-bearing years may extend into the sixties; people may routinely continue working into their eighties or beyond. With this expanded timeline will come all kinds of new possibilities, including vastly greater periods of intellectual creativity and a softening of the irrational behaviors that arise from the universal fear of death.

- I am optimistic that the longer life of the body will be accompanied by enhanced powers of the brain. We already live in world where it is getting harder and harder to forget. A simple Google search often retrieves long-lost trivia, historical experiences, even the names of long-dead relatives. What we have today is but a tiny taste of what lies ahead. Computing power is now so cheap, and wireless communication so effortless, that a person could easily wear a microphone (or even a low-resolution video camera) at all times and compile a digital database of every word he or she uttered.

  In the future, many people will choose to do so; we will all have personalized, searchable databases at our commands. Rapid advances in brain prostheses mean that soon we will be able to access those databases simply by the power of thought. Within a couple of decades, the information will be beamed back in a form the brain can interpret – we will be able to hear the playback much as deaf people can now hear the world with cochlear implants. Vision is slightly more difficult, but it, too, will eventually be reverse engineered. That will undoubtedly give space exploration a tremendous boost. Earthbound scientists will be able to 'inhabit' robotic explorers on other worlds, and any interested participant will be able to log on to experience the adventure. Humans will venture into space physically as well, but at first that will happen primarily for sport, I expect.

- I am optimistic that researchers, aided by their computers and longer careers, will crack the great twin mysteries of physics: the nature of gravity and the possibility of other dimensions. Here I am talking not just about theoretical advances (such as may occur at the Large Hadron Collider after it revs up late this year) that could bolster the theory that gravity, unlike the other forces, has the ability to transmit out of the three dimensions of human experience. I am also talking about a kookier optimism that our discoveries will have practical consequences. It may be possible to build instruments that can sense universes lying outside our three dimensions. It may be possible to manipulate gravity, turning it down where convenient (to launch a rocket, for instance) and cranking it up where desired (in a lab or on a space station). It may even be possible, as MIT cosmologist Alan Guth once speculated, to create a new universe as a laboratory experiment – the ultimate empirical investigation of the Big Bang.

- Finally, I am optimistic that with all of these intellectual and material achievements will come a science-based spiritual awakening. Back in the 1930s, Albert Einstein spoke of a 'cosmic religious feeling' and tried to convince the public (with painfully little success) that scientists are every bit as spiritual as are the world's religious leaders. It may not look that way now, but I think Einstein will soon be vindicated. Longer, more connected lives will eat away at the religion of fear, the rudimentary form of faith rooted in anxiety about loneliness and the apparent absoluteness of death.

More important, the next round of scientific discoveries promises a powerful new sense of our connection to the rest of the universe, and even to universes beyond our own. One of the most potent criticisms of science is that, unlike religion, it offers no sense of purpose. That has never been true – What greater purpose is there than intellectual exploration, the key trait distinguishing us from the other animals? – but now, more than ever, science has a chance to make its case. To do so, though, it needs to adapt. It needs to develop more of a communal structure. It needs to develop a humane language, expressing its findings explicitly as triumphs of human achievement. It needs to celebrate our ever expanding dominion over nature, while articulating a humble appreciation that nature is, indeed, where we all came from.

Above all, science needs a face, a representative (or representatives) as charismatic as Pope John Paul II or, say, the late Carl Sagan. Right now, the faces of science are selected by book sales, television specials, and pure self-promotion; its elected leaders, the heads of scientific societies, rarely function as public figures. Surely there is a better way, and the future will be all the better for it.

INDEX